Farming Human Pathogens

Rodrick Wallace · Deborah Wallace ·
Robert G. Wallace

Farming Human Pathogens

Ecological Resilience
and Evolutionary Process

 Springer

Rodrick Wallace
New York State Psychiatric Institute
Division of Epidemiology
Research Department
1051 Riverside Dr.
New York NY 10032
USA
wallace@pi.cpmc.columbia.edu

Deborah Wallace
549 West 123rd Street
New York NY 10032
Apt. 16F
USA
rdwall@ix.netcom.com

Robert G. Wallace
Department of Geography
University of Minnesota
3340 16th Avenue South
Minneapolis MN 55407
USA
rwallace24@gmail.com

ISBN 978-1-4419-2826-9   e-ISBN 978-0-387-92213-3
DOI 10.1007/978-0-387-92213-3
Springer Dordrecht Heidelberg London New York

Printed on acid-free paper

Springer is part of Springer Science+Business Media (www.springer.com)

Preface

Disease interventions, at both the individual and population levels, are, with a few bright exceptions, faltering. Vaccines, pharmaceuticals, and low-tech solutions, such as bed nets and water filters, while successful in addressing many reductionist diseases, cannot contain pathogens that use interactions at one level of biocultural organization to evolve out from underneath interventions directed at them at another. Such holistic diseases, operating across fluctuating swaths of space and time, infect and kill millions annually. HIV, tuberculosis, malaria, and influenza, among others, confound even the most concerted efforts.

Virologists, epidemiologists, evolutionary ecologists, population health geographers, drug designers, and public health officials must return to basic principles. Lab, field, and statistical apparatuses, powered now by industrial computing, appear inadequate to the task of rolling back many scourges old and newly emergent. New ways of thinking about basic biology, evolution, and scientific practice are in order. In a world in which viruses and bacteria evolve in response to humanity's multifaceted infrastructure – agricultural, transportation, pharmaceutical, public health, scientific, political – our epistemological and epidemiological intractabilities may be in fundamental ways one and the same. Some pathogens evolve into population states in which we cannot or, worse, refuse to think (Wallace and Wallace 2004).

In an attempt to break the current stalemate, we offer here the possibility that shifts in mesoscale ecosystem resilience can entrain punctuations in molecular cognition and gene expression on more rapid time scales. Through such means evolution by selection can also trigger transitions by punctuated equilibria across a variety of time scales.

The analysis, based on the computational systems biology approach of Wallace and Wallace (2008), reduces ecosystem, gene expression, and Darwinian inheritance to a least common denominator: information sources interacting by crosstalk at markedly different rates. Pettini's (2007) 'topological hypothesis', via a homology between information source uncertainty and free energy density, generates a regression-like class of statistical models of sudden coevolutionary phase transition based on the asymptotic limit theorems of information theory linking all three levels. A mathematical restatement of Holling's (1992) extended keystone hypothesis about the roles mesoscale phenomena play in entraining both slower and faster dynamical structures – mesoscale resonance – produces the key results. Into this informational turn we incorporate a cognitive paradigm for gene expression and ontogeny, mirroring I. R. Cohen's (2000) treatment of immune function, Gilbert's (2000, 2001, 2005) evo-devo perspective, and West-Eberhard's (2003, 2005) work on developmental plasticity and the origins of phylogenetic diversity.

In essence, the asymptotic limit theorems of communication theory impose necessary conditions on cognitive gene expression and its interaction with the embedding ecosystem and ontogeny in much the same way the central limit theorem imposes necessary conditions leading to the construction of regression models. It should be possible to fit the resulting statistical models to real data, providing new means of comparing particular organismal and epidemiological systems under different conditions, or different systems under similar conditions. Ultimately we offer new tools for the analysis of currently intractable phenomena involving broadly cognitive processes taking place within nested hierarchies of complex biochemical and socioecological networks.

A first application takes us to prebiotica. We show Eigen's paradox, tracing the transition from high error-low energy replication to some high energy-low error form, appears to be characterized by a fundamental protoecosystemic shift in metabolic resilience that entrained reproductive fidelity.

Our second, and central, application requires invoking the influence of humanity's cultural structures on pathogen ecosystems. We reconsider the evolutionary ecology of HIV, avian influenza H5N1, and other highly adaptable disease organisms. In particular, we examine how public policy and socioeconomic structure not only exert selection pressures on infectious diseases, but can actually 'farm' them in a broadly coevolutionary process driven both by reductionist medical interventions and by economically induced expansions in the pathogens' ecological niches. This leads us to call for an 'integrated pathogen management' similar to the 'integrated pest management' strategies increasingly advocated in agriculture. An essential feature of any pathogen management strategy, of course, would be widespread social and economic reform. Absent such intervention, evolutionarily responsive pathogens will in all likelihood continue to inundate us like hurricanes lined up in the Caribbean, leaving our population centers devastated in their terrible wake.

Contents

1

Introduction

Evolutionary biologists have long debated whether speciation occurs gradually or in response to massive catastrophes (Mayr 1982, Gould 2002). By mid-twentieth century the gradualists, buttressed by a mathematical synthesis of Darwinian selection and Mendelian genetics, prevailed, and the catastrophists, of various stripes, were professionally marginalized, going the way of the dinosaur. Speciation was now viewed as resulting from selectionist microevolution in response to incremental environmental challenges, a view best embodied by early efforts in ecological and population genetics.

The paradigm shift had the curious effects of alienating organism from environment, with evolution limited to changes in allelic frequencies (Lewontin, 2000). Concomitantly, evolution, outside cosmology the most historical of sciences, became curiously dehistoricized. There appeared an inability or refusal to incorporate evolutionary history into the mathematical mechanics tracing population dynamics beyond differentiable functions and Markov chains. History became largely relegated to the internal state of the system at the previous interval. The formalisms and their analytic outgrowths locked field studies and experimental programs into the most extreme versions of the adaptationist program.

By the mid-70s, however, a rearguard action emerged. Eldredge and Gould (1972; Gould and Eldredge 1977) concluded that speciation could occur, and in some systems often occurred, suddenly. Species appeared abruptly in the fossil record, remained largely unchanged in it, and then disappeared 'suddenly' on a geological time scale. These 'punctuated equilibria' appeared to integrate gradualism and catastrophism. In something of an epistemological resuscitation, a panoply of other potential causes complicating organismal adaptation by natural selection reemerged in the literature: laws of form a la D'Arcy Thompson, Galton's polyhedron, phylogenetic constraints, the historical contingencies of catastrophic climatic events and mass extinction, and selection at other levels of biological organization (including macroevolutionary species selection).

R. Wallace et al., *Farming Human Pathogens*, DOI 10.1007/978-0-387-92213-3_1,
© Springer Science+Business Media, LLC 2009

In the same period, C. S. Holling (1973) developed the ecological equivalent of punctuated equilibrium. Resilience theory (Fleming and Shoemaker, 1992; Gunderson 2000; Volney and Fleming 2007) views each ecosystem, formed by the interactions among community species and with their environment, as normally in a state of quasi-equilibrium. As the system is subjected to various impacts, it shows no obvious change in structure or function, but the relationships among many resident species become tighter as the perturbations erode the delicate peripheral interactions. Finally, either a particularly intense impact on the ecosystem or the aggregated effect of multiple impacts degrades so many of the relationships that those that remain shatter. The ecosystem shifts relatively suddenly into a different dynamic domain – a different quasi-equilibrium with markedly different structure and function. A forest shifts into prairie land after long-term drought. With a change in local hunting practices, an epizootic erupts out from a state of endemism. Eutrophication emerges from enough agricultural runoff and urban wastewater.

While the analogy has long been apparent (e.g., Levin 1999), we attempt here to more forcefully engage the possibility Holling's theory provides something of an explanation for punctuated equilibrium. If ecosystems are suddenly transformed into different configurations, species are confronted with sudden changes in selection pressure. Along with the fossil record, climatological and geological data show major changes in temperature, atmospheric composition, and shifts in volcanism, earthquakes, and tectonic plates. These form the macroscale of ecosystems. In the other direction, local topography, geology, hydrology, and microclimate offer abiotic contributions to microscale ecological niches. By their activities organisms also modify their own niches and those of other organisms (Lewontin, 1993; Odling-Smee et al., 1996, 2003). Sandwiched between, and at times overlapping, landscape processes, such as wildfires spreading over large numbers of niches, form the mesoscale (Holling, 1992).

Given the geographically defined and conditional nature of the environment to which organisms respond, not all organismal genes are expressed at any single time (above and beyond the effects of age-specific ontogeny). The genetic variation within a local population may be greater than the phenotype presented in the field (Gibson and Dworkin, 2004). The niche may select for a phenotype consonant with the expression of only a limited genetic combination held within a cryptic reservoir. A domain shift in the ecological framework could lead to selection for qualitatively different phenotypes, a la Galton's polyhedron. If the ecological shifts extend beyond what available phenotypic plasticity can buffer, those lineages with the genetic resources needed to express the newly preferred phenotypes, perhaps via a kind of neo-orthogenetic archive, will by way of classic selection supplant standing morphs. As the new ecological configuration hardens, the genetic composition of the new clado-morphs may include allelic combinations that contribute to reinforcing isolation mechanisms and speciation (West-Eberhard, 2003, 2005). By another mechanism – the Baldwin effect – behavioral traditions differentiating populations can become canalized into the genome (Avital and Jablonka, 2000; Weber and Depew, 2003). In sum, a variety of embedding conditions can help entrain an eco-evo-devo composite object.

Here we will analyze the interaction of these phenomena using a principled approach which reduces ecosystem dynamics, gene expression, and genetic inheritance via natural selection to a least common denominator of interacting information sources constrained by the asymptotic limit theorems of communications theory. This is not an entirely new perspective. Priami (2007), for example, finds that the interaction between biological entities can be represented as an exchange of information between programs. Earlier, Waddington (1972; Jimenez-Montano, 1989) had suggested that language may become a paradigm for a theory of general biology, but a language in which basic sentences are programs, not simple statements. Our contribution is to hew very closely to the basic mathematical structure of the asymptotic limit theorems of information theory and the associated generalizations afforded by the large deviations program of applied probability.

We will begin by examining the interaction between our three information sources: ecosystem dynamics, gene expression, and genetic inheritance. We will use the homology between information source uncertainty and the free density of a physical system to import phase transition methods from statistical physics via Pettini's (2007) topological hypothesis. An analog to the Onsager relations of nonequilibrium statistical mechanics permits studying these systems far from phase transition, leading to a coevolutionary paradigm. The reexpression of Ancel's (1999) work on the Baldwin effect in terms of a 'tuning theorem' variant of the Shannon coding theorem produces the essential result that mesoscale ecosystem shifts are likely particularly powerful in entraining gene expression and evolutionary selection.

We then apply these perspectives to a number of fundamental problems in biology and health, beginning with Eigen's paradox in prebiotic evolution. Extending the argument permits modeling the effects of embedding cultural structures. The formal machinery provides insights about infectious disease dynamics in human ecosystems. In particular, we study the roles played by reductionist interventions, regressive public policy, and rapacious economic practice in 'farming' human pathogens.

1.1 Ecosystems as information sources

1.1.1 Coarse-graining a simple model

We begin with a simplistic picture of an elementary predator/prey ecosystem which, nonetheless, provides a useful pedagogical starting point. Let X represent the appropriately scaled number of predators, Y the scaled number of prey, t the time, and ω a parameter defining their interaction. The model assumes that the ecologically dominant relation is an interaction between predator and prey, so that

$$dX/dt = \omega Y$$

$$dY/dt = -\omega X$$

(1.1)

Thus the predator populations grows proportionately to the prey population, and the prey declines proportionately to the predator population.

After differentiating the first and using the second equation, we obtain the simple relation

$$d^2 X/dt^2 + \omega^2 X = 0$$

(1.2)

having the solution

$$X(t) = sin(\omega t); Y(t) = cos(\omega t).$$

with

$$X(t)^2 + Y(t)^2 = sin^2(\omega t) + cos^2(\omega t) \equiv 1.$$

Thus in the two dimensional phase space defined by $X(t)$ and $Y(t)$, the system traces out an endless, circular trajectory in time, representing the out-of-phase sinusoidal oscillations of the predator and prey populations.

Divide the $X - Y$ phase space into two components – the simplest coarse graining – calling the halfplane to the left of the vertical Y-axis A and that to the right B. This system, over units of the period $1/(2\pi\omega)$, traces out a stream of A's and B's having a single very precise grammar and syntax:

$$ABABABAB...$$

Many other such statements might be conceivable, e.g.,

$$AAAAA..., BBBBB..., AAABAAAB..., ABAABAAAB...,$$

and so on, but, of the obviously infinite number of possibilities, only one is actually observed, is 'grammatical': $ABABABAB....$

More complex dynamical system models, incorporating diffusional drift around deterministic solutions, or even very elaborate systems of complicated stochastic differential equations, having various domains of attraction, that is, different sets of grammars, can be described by analogous symbolic dynamics (Beck and Schlogl, 1993, Ch. 3).

1.1.2 Ecosystems and information

Rather than taking symbolic dynamics as a simplification of more exact analytic or stochastic approaches, it proves useful, as it were, to throw out the Cheshire cat but keep the cat's smile, generalizing symbolic dynamics to a more comprehensive information dynamics. Ecosystems may not have identifiable sets of stochastic dynamic equations like noisy, nonlinear mechanical clocks, but, under appropriate coarse-graining, they may still have recognizable sets of grammar and syntax over the long-term.

Examples abound. The turn-of-the seasons in a temperate climate, for many natural communities, looks remarkably the same year after year: the ice melts, the migrating birds return, the trees bud, the grass grows, plants and animals reproduce, high summer arrives, the foliage turns, the birds leave, frost, snow, the rivers freeze, and so on.

Suppose it is indeed possible to empirically characterize an ecosystem at a given time t by observations of both habitat parameters such as temperature and rainfall, and numbers of various plant and animal species.

Traditionally, one can then calculate a cross-sectional species diversity index at time t using an information or entropy metric of the form

$$H = -\sum_{j=1}^{M}(n_j/N)\log[(n_j/N)],$$

$$N \equiv \sum_{j=1}^{M} n_j$$

where n_j is the number of observed individuals of species j and N is the total number of individuals of all species observed (e.g., Pielou, 1977; Ricotta, 2003; Fath et al., 2003).

This is not the approach taken here. Quite the contrary, in fact. Suppose it is possible to coarse grain the ecosystem at time t according to some appropriate partition of the phase space in which each division A_j represent a particular range of numbers of each possible species in the ecosystem, along with associated parameters such as temperature, rainfall, and the like. What is of particular interest to our development is not cross sectional structure, but rather longitudinal paths, that is, ecosystem statements of the form $x(n) = A_0, A_1, ..., A_n$ defined in terms of some natural time unit of the system. Thus n corresponds to an again appropriate characteristic time unit T, so that $t = T, 2T, ..., nT$.

To reiterate, unlike the traditional use of information theory in ecology, our interest is in the *serial correlations along paths*, and not at all in the cross-sectional entropy calculated for of a single element of a path.

Let $N(n)$ be the number of possible paths of length n which are consistent with the underlying grammar and syntax of the appropriately coarsegrained ecosystem: spring leads to summer, autumn, winter, back to spring, etc., but never something of the form spring to autumn to summer to winter in a temperate ecosystem.

The fundamental assumptions are that – for this chosen coarse-graining – $N(n)$, the number of possible grammatical paths, is much smaller than the total number of paths possible, and that, in the limit of (relatively) large n,

$$H = \lim_{n \to \infty} \frac{\log[N(n)]}{n}$$

(1.3)

both exists and is independent of path.

This is a critical foundation to, and limitation on, the modeling strategy and its range of strict applicability, but is, in a sense, fairly general since it is *independent of the details of the serial correlations along a path*.

Again, these conditions are the essence of the parallel with parametric statistics. Systems for which the assumptions are not true will require special nonparametric approaches. We are inclined to believe, however, that, as for parametric statistical inference, the methodology will prove robust in that many systems will sufficiently fulfill the essential criteria.

This being said, not all possible ecosystem coarse-grainings are likely to work, and different such divisions, even when appropriate, might well lead to different descriptive quasi-languages for the ecosystem of interest. The example of Markov models is relevant. The essential Markov assumption is that the probability of a transition from one state at time T to another at time $T + \Delta T$ depends only on the state at T, and not at all on the history by which that state was reached. If changes within the interval of length ΔT are plastic, or path dependent, then attempts to model the system as a Markov process *within* the natural interval ΔT will fail, even though the model works quite well for phenomena separated by natural intervals.

Thus empirical identification of relevant coarse-grainings for which this body of theory will work is clearly not trivial, and may, in fact, constitute the hard scientific core of the matter.

This is not, however, a new difficulty in ecosystem theory. Holling (1992), for example, explores the linkage of ecosystems across scales, finding that mesoscale structures – what might correspond to the neighborhood in a human community – are ecological keystones in space, time, and population, which drive process and pattern at both smaller and larger scales and levels of organization. This will, in fact, be a core argument of our development which we will formally derive in chapter 2 using an analog to the 'no free lunch' theorem of computational optimization theory (English, 1996).

Levin (1989) argues that there is no single correct scale of observation: the insights from any investigation are contingent on the choice of scales. Pattern is neither a property of the system alone nor of the observer, but of an interaction between them. Pattern exists at all levels and at all scales, and recognition of this multiplicity of scales is fundamental to describing and understanding ecosystems. In his view there can be no 'correct' level of aggregation: we must recognize explicitly the multiplicity of scales within ecosystems, and develop a perspective that looks across scales and that builds on a multiplicity of models rather than seeking the single 'correct' one.

Given an appropriately chosen coarse-graining, whose selection in many cases will be the difficult and central trick of scientific art, suppose it possible to define joint and conditional probabilities for different ecosystem paths, having the form $P(A_0, A_1, ..., A_n), P(A_n|A_0, ..., A_{n-1})$, such that appropriate joint and conditional Shannon uncertainties can be defined on them. For paths of length two these would be of the form

$$H(X_1, X_2) \equiv -\sum_j \sum_k P(A_j, A_k) \log[P(A_j, A_k)]$$

$$H(X_1|X_2) \equiv -\sum_j \sum_k P(A_j, A_k) \log[P(A_j|A_k)],$$

(1.4)

where the X_j represent the stochastic processes generating the respective paths of interest.

The essential content of the Shannon-McMillan Theorem is that, for a large class of systems characterized as information sources, a kind of law-of-large numbers exists in the limit of very long paths, so that

$$H[X] = \lim_{n \to \infty} \frac{\log[N(n)]}{n} =$$

$$\lim_{n \to \infty} H(X_n|X_0, ..., X_{n-1}) =$$

$$\lim_{n \to \infty} \frac{H(X_0, X_1, ..., X_n)}{n+1}.$$

(1.5)

Taking the definitions of Shannon uncertainties as above, and arguing backwards from the latter two equations (Khinchin, 1957), it is indeed possible to recover the first, and divide the set of all possible temporal paths of our ecosystem into two subsets, one very small, containing the grammatically correct, and hence highly probable paths, which we will call 'meaningful', and a much larger set of vanishingly low probability.

Basic material on information theory can be found in any number of texts, for example, Ash (1990), Khinchin (1957), Cover and Thomas (1991).

The next task is to show how the cognitive processes which so distinguish much individual and collective animal activity, as well as many basic physiological processes, can be fitted into a similar context, that is, characterized as information sources.

1.2 Cognition as an information source

Atlan and Cohen (1998) argue that the essence of cognition is comparison of a perceived external signal with an internal, learned picture of the world, and then, upon that comparison, the choice of one response from a much larger repertoire of possible responses. Such reduction in uncertainty inherently carries information, and, following the approach of Wallace (2000), or Wallace and Fullilove (2008), it is possible to make a very general model of this process as an information source.

Cognitive pattern recognition-and-selected response, as conceived here, proceeds by convoluting an incoming external 'sensory' signal with an internal 'ongoing activity' – which includes, but is not limited to, the learned picture of the world – and, at some point, triggering an appropriate action based on a decision that the pattern of sensory activity requires a response. It is not necessary to specify how the pattern recognition system is trained, and hence possible to adopt a weak model, regardless of learning paradigm, which can itself be more formally described by the Rate Distortion Theorem. Fulfilling Atlan and Cohen's (1998) criterion of meaning-from-response, we define a language's contextual meaning entirely in terms of system output.

The model, an extension of that presented in Wallace (2000), is as follows.

A pattern of 'sensory' input, say an ordered sequence $y_0, y_1, ...$, is mixed in a systematic (but unspecified) algorithmic manner with internal 'ongoing' activity, the sequence $w_0, w_1, ...$, to create a path of composite signals $x = a_0, a_1, ..., a_n, ...$, where $a_j = f(y_j, w_j)$ for some function f. This path is then fed into a highly nonlinear, but otherwise similarly unspecified, decision oscillator which generates an output $h(x)$ that is an element of one of two (presumably) disjoint sets B_0 and B_1. We take

$$B_0 \equiv b_0, ..., b_k,$$

$$B_1 \equiv b_{k+1}, ..., b_m.$$

(1.6)

Thus we permit a graded response, supposing that if

$$h(x) \in B_0$$

(1.7)

the pattern is not recognized, and if

$$h(x) \in B_1$$

(1.8)

the pattern is recognized and some action $b_j, k + 1 \leq j \leq m$ takes place.

Our focus is on those composite paths x that trigger pattern recognition-and-response. That is, given a fixed initial state a_0, such that $h(a_0) \in B_0$, we examine all possible subsequent paths x beginning with a_0 and leading to the event $h(x) \in B_1$. Thus $h(a_0, ..., a_j) \in B_0$ for all $0 \le j < m$, but $h(a_0, ..., a_m) \in B_1$.

For each positive integer n, let $N(n)$ be the number of grammatical and syntactic high probability paths of length n which begin with some particular a_0 having $h(a_0) \in B_0$ and lead to the condition $h(x) \in B_1$. We shall call such paths meaningful and assume $N(n)$ to be considerably less than the number of all possible paths of length n – pattern recognition-and-response is comparatively rare. We again assume that the longitudinal finite limit $H \equiv \lim_{n \to \infty} \log[N(n)]/n$ both exists and is independent of the path x. We will – not surprisingly – call such a cognitive process *ergodic*.

Note that disjoint partition of state space may be possible according to sets of states which can be connected by meaningful paths from a particular base point, leading to a natural coset algebra of the system, a groupoid. This is a matter of some importance pursued at length in the next chapter.

It is thus possible to define an ergodic information source $\mathbf{X}$ associated with stochastic variates X_j having joint and conditional probabilities $P(a_0, ..., a_n)$ and $P(a_n | a_0, ..., a_{n-1})$ such that appropriate joint and conditional Shannon uncertainties may be defined which satisfy the relations of equation (1.5) above.

This information source is taken as *dual* to the ergodic cognitive process.

We reiterate that the Shannon-McMillan Theorem and its variants provide 'laws of large numbers' which permit definition of the Shannon uncertainties in terms of cross-sectional sums of the form $H = -\sum P_k \log[P_k]$, where the P_k constitute a probability distribution.

It is important to recognize that different quasi-languages will be defined by different divisions of the total universe of possible responses into various pairs of sets B_0 and B_1. Like the use of different distortion measures in the Rate Distortion Theorem (e.g., Cover and Thomas, 1991), however, it seems obvious that the underlying dynamics will all be qualitatively similar.

Nonetheless, dividing the full set of possible responses into the sets B_0 and B_1 may itself require higher order cognitive decisions by another module or modules, suggesting the necessity of choice within a more or less broad set of possible quasi-languages. This would directly reflect the need to shift gears according to the different challenges faced by the organism or social group. A critical problem then becomes the choice of a normal zero-mode language among a very large set of possible languages representing the excited states accessible to the system. This is a fundamental matter which mirrors, for isolated cognitive systems, the resilience arguments applicable to more conventional ecosystems, that is, the possibility of more than one zero state to a cognitive system. Identification of an excited state as the zero mode becomes, then, a kind of generalized autoimmune disorder which can be triggered by linkage with external ecological information sources representing various kinds of structured psychosocial stress, a matter we explore at length elsewhere (Wallace and Fullilove, 2008; Wallace, 2008b).

In sum, meaningful paths – creating an inherent grammar and syntax – have been defined entirely in terms of system response, as Atlan and Cohen (1998) propose.

This formalism can easily be applied to the stochastic neuron in a neural network, as done in Wallace (2005a).

Ultimately it becomes necessary to parametize the information source uncertainty of the dual information source to a cognitive pattern recognition-and-response with respect to one or more variates, writing $H[\mathbf{K}]$, where $\mathbf{K} \equiv (K_1, ..., K_s)$ represents a vector in a parameter space. Let the vector $\mathbf{K}$ follow some path in time, that is, trace out a generalized line or surface $\mathbf{K}(t)$. We assume that the probabilities defining H, for the most part, closely track changes in $\mathbf{K}(t)$, so that along a particular piece of a path in parameter space the information source remains as close to stationary – the probabilities are fixed in time – and ergodic as is needed for the mathematics to work. Such a system is characterized as 'adiabatic' in the physics literature. Between pieces we will, below, impose phase transition characterized by a renormalization symmetry, in the sense of Wilson (1971), as done in chapter 3.

Such an information source will be termed 'adiabatically piecewise stationary ergodic' (APSE).

To reiterate, the ergodic nature of the information sources is a generalization of the law of large numbers and implies that the long-time averages we will need to calculate can, in fact, be closely approximated by averages across the probability spaces of those sources. For non-ergodic information sources, a function, $\mathcal{H}(x_n)$, of each path $x_n \to x$, may be defined, such that $\lim_{n \to \infty} \mathcal{H}(x_n) = \mathcal{H}(x)$, but $\mathcal{H}$ will not in general be given by the simple cross-sectional laws-of-large numbers analogs above (Khinchin, 1957).

Let $s \equiv d(x, \hat{x})$ for high probability paths x and $\hat{x}$, where d is a distortion measure, as described in the Appendix. For 'nearly' ergodic systems one might use something of the form

$$\mathcal{H}(\hat{x}) \approx \mathcal{H}(x) + sd\mathcal{H}/ds|_{s=0}$$

for s sufficiently small. Loosely speaking, the idea is to take a distortion measure as a kind of Finsler metric, imposing a resulting 'global' structure over an appropriate class of non-ergodic information sources. One possible interesting theorem, then, obviously revolves around what properties are metric-independent, in much the same manner as the Rate Distortion Theorem.

This heuristic sketch can be made more precise as follows:

Take a set of 'high probability' paths $x_n \to x$.

Suppose, for all such x, there is an open set, U, containing x, on which the following conditions hold:

(i) For all paths $\hat{x}_n \to \hat{x} \in U$, a distortion measure $s_n \equiv d_U(x_n, \hat{x}_n)$ exists.

(ii) For each path $x_n \to x$ in U there exists a pathwise invariant function $\mathcal{H}(x_n) \to \mathcal{H}(x)$, in the sense of Khinchin (1957, p.72). While such a function will almost always exist, only in the case of an ergodic information source can it be identified as an 'entropy' in the usual sense.

(iii) A function $F_U(s_n, n) \equiv f_n \to f$ exists, for example,

$$f_n = s_n, \log[s_n]/n, s_n/n,$$

and so on.

(iv) The limit

$$\lim_{n \to \infty} \frac{\mathcal{H}(x_n) - \mathcal{H}(\hat{x}_n)}{f_n} \equiv \nabla_F \mathcal{H}|_x$$

exists and is finite.

Under such conditions, various nontrivial standard global atlas/manifold constructions are possible. Again, $\mathcal{H}$ is not simply given by the usual expressions for source uncertainty if the source is not ergodic, and the phase transition development of subsequent chapters may be correspondingly more complicated. Restriction to high probability paths simplifies matters considerably, although precisely characterizing them may be difficult, requiring extension of the Shannon-McMillan Theorem and its Rate Distortion generalization.

The essential unanswered question is under what circumstances the differential treatment above for 'almost' ergodic information sources permits something very much like what Khinchin (1957, p. 54) calls the 'E property' enabling classification of paths into a small set of high probability and a vastly larger set of vanishingly small probability (Khinchin, 1957, p. 74).

This development has close parallels with Dretske's (1981, 1988, 1992, 1994) speculations on the the role of the asymptotic limit theorems of information theory in constraining high level mental function.

Wallace (2004, 2005a) and Wallace and Fullilove (2008) describe in some detail how, for larger animals, immune function, tumor control, the hypothalamic pituitary adrenal (HPA) axis (the flight-or-fight system), emotion, conscious thought, and embedding group (and sometimes cultural) structures are all cognitive in this simple sense. In general these cognitive phenomena will occur at far faster rates than embedding ecosystem changes.

It is worth a more detailed recounting of the arguments for characterizing a number of physiological subsystems as cognitive from the viewpoint of this section.

1.2.1 Immune cognition

Atlan and Cohen (1998) have proposed an information-theoretic cognitive model of immune function and process, a paradigm incorporating cognitive pattern recognition-and-response behaviors analogous to those of the central nervous system. This work follows in a very long tradition of speculation on the cognitive properties of the immune system (e.g., Tauber, 1998; Podolsky and Tauber, 1998; Grossman, 1989, 2000).

From the Atlan/Cohen perspective, the meaning of an antigen can be reduced to the type of response the antigen generates. That is, the meaning of an antigen is functionally defined by the response of the immune system. The meaning of an antigen to the system is discernible in the type of immune response produced, not merely whether or not the antigen is perceived by the receptor repertoire. Because the meaning is defined by the type of response there is indeed a response repertoire and not only a receptor repertoire.

To account for immune interpretation Cohen (1992, 2000) has reformulated the cognitive paradigm for the immune system. The immune system can respond to a given antigen in various ways, it has 'options.' Thus the particular response we observe is the outcome of internal processes of weighing and integrating information about the antigen.

In contrast to Burnet's view of the immune response as a simple reflex, it is seen to exercise cognition by the interpolation of a level of information processing between the antigen stimulus and the immune response. A cognitive immune system organizes the information borne by the antigen stimulus within a given context and creates a format suitable for internal processing. The antigen and its context are transcribed internally into the 'chemical language' of the immune system.

The cognitive paradigm suggests a language metaphor to describe immune communication by a string of chemical signals. This metaphor is apt because the human and immune languages can be seen to manifest several similarities such as syntax and abstraction. Syntax, for example, enhances both linguistic and immune meaning.

Although individual words and even letters can have their own meanings, an unconnected subject or an unconnected predicate will tend to mean less than does the sentence generated by their connection.

The immune system creates a 'language' by linking two ontogenetically different classes of molecules in a syntactical fashion. One class of molecules are the T and B cell receptors for antigens. These molecules are not inherited, but are somatically generated in each individual. The other class of molecules responsible for internal information processing is encoded in the individual's germline.

Meaning, the chosen type of immune response, is the outcome of the concrete connection between the antigen subject and the germline predicate signals.

The transcription of the antigens into processed peptides embedded in a context of germline ancillary signals constitutes the functional 'language' of the immune system. Despite the logic of clonal selection, the immune system does not respond to antigens as they are, but to abstractions of antigens-in-context. Cohen (2006) provides a more recent perspective, focusing on inflammatory processes as maintenance in which the immune decision-making process uses strategies similar to those observed in the nervous system.

1.2.2 Tumor control

We argue that the next larger cognitive submodule after the immune system must be a tumor control mechanism that may include immune surveillance, but clearly transcends it. Nunney (1999) has explored cancer occurrence as a function of animal size, suggesting that in larger animals, whose lifespan grows as about the 4/10 power of their cell count, prevention of cancer in rapidly proliferating tissues becomes more difficult in proportion to size. Cancer control requires the development of additional mechanisms and systems to address tumorigenesis as body size increases – a synergistic effect of cell number and organism longevity. Nunney (1999, p. 497) concludes that this pattern may represent a real barrier to the evolution of large, long-lived animals and predicts that those that do evolve have recruited additional controls over those of smaller animals to prevent cancer.

Different tissues may have evolved markedly different tumor control strategies. All of these, however, are likely to be energetically expensive, permeated with different complex signaling strategies, and subject to a multiplicity of reactions to signals, including those related to psychosocial stress. Forlenza and Baum (2000) explore the effects of stress on the full spectrum of tumor control in higher animals, ranging from DNA damage and control, to apoptosis, immune surveillance, and mutation rate. Elsewhere (R. Wallace et al., 2003) we argue that this elaborate tumor control strategy, particularly in large animals, must be at least as cognitive as the immune system itself, which is one of its components. That is, some comparison must be made with an internal picture of a 'healthy' cell, and a choice made as to response: none, attempt DNA repair, trigger programmed cell death, engage in full-blown immune attack. This is, from the Atlan/Cohen perspective, the essence of cognition.

1.2.3 A cognitive paradigm for gene expression

Modes of genetic inheritance are assumed well understood since evolutionary theory's Modern Synthesis and the discovery of DNA and its translation. But the mechanisms of gene activation, regulation, and expression remain largely hidden. A broad reading of the literature illuminates a stark and increasingly mysterious landscape.

Liu and Ringner (2007) find gene expression signatures consisting of tens to hundreds of genes determine different biological states and conclude that it is crucial to systematically analyze gene expression signatures in the context of signaling pathways.

Soyer et al. (2006) find that, although massive network structures are associated with the biological signal transduction allowing a cell or organism to sense its environment and react accordingly, the experimental work needed to gather enough quantitative information to develop accurate mathematical models is highly labor intensive. The modeling of specific networks, then, may be of limited use in developing a broad understand of the general properties of biological signaling networks.

One possible mathematical characterization of these difficulties is found in Sayyed-Ahmad et al. (2007), who explore the basic conundrum in terms of a dynamic model. In their view the state of a cell is specified by a set of variables Ψ for which we know the governing equations and a set T which is at the frontier of our understanding (that is, for which we do not know the governing equations). The challenge is that the dynamics of Ψ is given by a cell model,

$$d\Psi/dt = G(\Psi, T(t), \Lambda)$$

(1.9)

in which the rate G depends not only on many rate and equilibrium constants Λ, but also on the time-dependent frontier variables $T(t)$. The descriptive variables, Ψ, can only be determined as a function of the unknown time courses $T(t)$. Thus the model cannot be simulated.

Liao et al (2003) find that using statistical methods on biological networks, such as principal component analysis, ignores the underlying network structures. The decompositions are based on a priori statistical constraints on the computed component signals. Such decomposition, in their view, provides a phenomenological model for the observed data and does not necessarily contain physically or biologically meaningful signals.

Baker and Stock (2007), however, pose the questions in a more general manner. They apply an information metaphor in which the understanding of signal transduction systems has focused on mechanisms that allow crosstalk between different information processing modalities. They ask about the decision-making mechanisms by which a bacterium controls the activities of its genes and proteins to adapt to changing environmental conditions. That is, how is information converted into knowledge, and how is knowledge sorted, evaluated and combined to guide action, morphogenesis and growth?

O'Nuallain (2006) provides an important perspective on this approach. In his view the categorical failure to solve the general problem of natural language processing by computer is prognostic of the future of gene expression work. After what seemed like a promising start, in his view, the field was stalled by an inability to handle, or even define coherently, 'contextual' factors. Currently, he continues, the field is gradually being taken over by Bayesian 'methods' that simply look for the statistical incidence of co-occurrence of lexical items in the source (analogous to gene) and target (analogous to protein) languages. Contextual factors in the case of gene expression include the bioenergetic status of the cell, a status that can be assessed properly only with painstaking work. And yet it determines what genes are turned on and off at any particular moment.

It seems clear that 18th Century dynamical models using 19th Century differential equation generalizations of equation (1.9) have little to offer in addressing fundamental questions of gene activation and regulation. More sophisticated work must clearly move in the direction of an Atlan/Cohen cognitive paradigm for gene expression, characterizing the processes, and their embedding contexts, in terms of nested sets of interacting dual information sources, whose behavior is constrained by the necessary conditions imposed by the asymptotic limit theorems of communications theory.

That is, properly coarse-grained and nested biochemical networks will have an observed grammar and syntax, and, constrained by powerful probability limit theorems, such description can enable construction of robust and meaningful statistical models of gene expression which may then be used for real scientific inference.

In sum, generalizing symbolic dynamics to a more inclusive, and less restrictive, cognitive paradigm for gene expression while invoking the inherent complexities of topological groupoids described in Wallace and Fullilove (2008) and Glazebrook and Wallace (2007), seems likely to provide badly needed illumination for this dark and confusing realm.

Not uncharacteristically, I.R. Cohen and colleagues (Cohen and Harel, 2007) have, in fact, already proposed something much in this direction, using a 'reactive system' paradigm for gene expression taken from computer models. Reactive systems, in their view, call our attention to their emergent properties. An emergent property of a system is a behavior of the system, taken as a whole, that is not expressed by any one of the lower-scale components that comprise it. Although Cohen and Harel (2007) then attempt to develop a complicated computer modeling strategy to address such reactive systems, Cohen (2006) describes in some detail the essential differences between such systems and conventional computer architecture. There is no external operator or programmer, no programs, algorithms or software distinct from the system's hardware, no central processing unit, no operating system, no formal mathematical logic, no termination criteria, since the system never stops, no verification procedures, and so on.

Zhu et al. (2007), by contrast, take an explicit kinetic chemical reaction approach to gene expression involving delayed stochastic differential equations. They begin by coarse-graining multi-step biochemical processes with single-step delayed reactions. Their coarse-graining involves not only collapsing biochemical steps, but collapsing the inevitable associated serial correlations into a small number of 'time delays'. The key feature of their model is that the complex multiple-step biochemical processes – as transcription, translation, and even the whole gene expression – are simplified to single-step time delayed reactions.

While there are sufficiently many gene expression mechanisms so that some of them, at least, will yield to this method, we are interested in those which are more complex, and indeed broadly cognitive, subject to emergent patterns which cannot be modeled simply as bifurcations of stochastically-perturbed mechanistic models.

Indeed, rather than pursuing the computer models that Cohen and Harel (2007) and Zhu et al. (2007) invoke, here we will attempt to extend our statistical and dynamic analytic treatment of the cognitive paradigm to a structure incorporating gene expression in a broadly coevolutionary manner. As Richard Hamming so famously put it, "The purpose of computing is insight, not numbers", and analytic models offer transparency as well as insight. We will, however, recover a phenomenological formalism as a kind of generalized Onsager model, but at a later, and far more global, stage of structure. That is, invocation of the necessary conditions imposed by the limit theorems of communication theory enables us to penetrate one layer deeper before it becomes necessary to call for an empirically-determined system of recursive stochastic differential equations.

This is not altogether a new perspective, although Cohen and Harel (2007) are perhaps the first to make an explicitly formal attack. Gilbert (2001) has put these matters as follows:

"Developmental plasticity (sometimes called phenotypic plasticity) is the notion that the genome enables the organism to produce a range of phenotypes. There is not a single phenotype produced by a particular genotype. The structural phenotype instructed by the environmental stimulation is referred to as a morph. When developmental plasticity manifests itself as a continuous spectrum of phenotypes expressed by a single genotype across a range of environmental conditions, this spectrum is called the norm of reaction... The reaction norm is thought to be a property of the genome and can also be selected [by evolutionary process]. Different genotypes will be expected to differ in the direction and amount of plasticity that they are able to express...

The environment is not merely a permissive factor in development. It can also be instructive. A particular environment can elicit different phenotypes from the same genotype. Development usually occurs in a rich environmental milieu, and most animals are sensitive to environmental cues. The environment may determine sexual phenotype, induce remarkable structural and chemical adaptations according to the season, induce specific morphological changes that allow an individual to escape predation... The environment can also alter the structure of our neurons and the specificity of our immunocompetent cells. We can give a definite answer to the question posed by Wolpert in 1994:

Will the egg be computable? That is, given a total description of the fertilized egg – the total DNA sequence and the location of all proteins and RNA – could one predict how the embryo will develop?

The answer has to be 'No. And thank goodness'. The phenotype depends to a significant degree on the environment, and this is a necessary condition for integrating the developing organism into its particular habitat."

West-Eberhard (2005) describes these dynamics in similar terms:

"Any new input, whether it comes from the genome, like a mutation, or from the external environment, like a temperature change, a pathogen, or a parental opinion, has a developmental effect only if the preexisting phenotype is responsive to it... A new input... causes a reorganization of the phenotype, or 'developmental recombination.'...In developmental recombination, phenotypic traits are expressed in new or distinctive combinations during ontogeny, or undergo correlated quantitative change in dimensions...Developmental recombination can result in evolutionary divergence... at all levels of organization.

Individual development can be visualized as a series of branching pathways. Each branch point is a developmental decision, or switch point, governed by some regulatory apparatus, and each switch point defines a modular trait. Developmental recombination implies the origin or deletion of a branch and a new or lost modular trait. It is important to realize that the novel regulatory response and the novel trait originate simultaneously. Their origins are, in fact, inseparable events: you cannot have a change in the phenotype, a novel phenotypic state, without an altered developmental pathway...

Contrary to the notion that mutational novelties have superior evolutionary potential, there are strong arguments for the greater evolutionary potential of environmental induced novelties. An environmental factor can affect numerous individuals, whereas a mutation initially can affect only one."

The phenomena West-Eberhard invokes can clearly be rephrased in terms of cognitive processes guided by the 'context' provided by the embedding environment, as we do in the formalism developed here.

1.3 Darwinian genetic inheritance as an information source

Adami et al. (2000) make a case for reinterpreting the Darwinian transmission of genetic heritage in terms of a formal information process. They assert that genomic complexity can be identified with the amount of information a sequence stores about its environment: genetic complexity can be defined in a consistent information-theoretic manner. In their view, information cannot exist in a vacuum and must be instantiated. For biological systems information is instantiated, in part, by DNA. To some extent it is the blueprint of an organism and thus information about its own structure. More specifically, it is a blueprint of how to build an organism that can best survive in its native environment, and pass on that information to its progeny. Adami et al. assert that an organism's DNA thus is not only a 'book' about the organism, but also a book about the environment it lives in, including the species with which it co-evolves. They identify the complexity of geonomes by the amount of information they encode about the world in which they have evolved.

Ofria et al. (2003) continue in the same direction and argue that genomic complexity can be defined rigorously within standard information theory as the information the genome of an organism contains about its environment. From the point of view of information theory, it is convenient to view Darwinian evolution on the molecular level as a collection of information transmission channels, subject to a number of constraints. In these channels, they state, the organism's genome codes for the information (a message) to be transmitted from progenitor to offspring, subject to noise from an imperfect replication process and multiple sources of contingency. Information theory is concerned with analyzing the properties of such channels, how much information can be transmitted and how the rate of perfect information transmission of such a channel can be maximized.

Adami and Cerf (2000) argue, using simple models of genetic structure, that the information content, or complexity, of a genomic string by itself (without referring to an environment) is a meaningless concept and a change in environment (catastrophic or otherwise) generally leads to a pathological reduction in complexity.

The transmission of genetic information is thus a contextual matter which involves operation of an information source which, according to this development, must interact with embedding (ecosystem) structures. Such interaction is, as we show in the next chapter, often highly punctuated, modulated by mesoscale ecosystem transitions via a generalization of the Baldwin effect akin to stochastic resonance, i.e., 'mesoscale resonance'.

2

Formal theory I

This chapter extends recent mathematical approaches to high level cognitive function and distributed cognition (Wallace 2005a; Wallace and Fullilove, 2008). It provides a general theory that can be applied to a cognitive paradigm for gene expression. Several such processes may, in fact, operate simultaneously or sequentially and interact with each other as well as with larger, embedding, structures: not one, but many, cognitive gene expression phenomena. Addressing this level of complexity requires significant mathematical overhead. Given a set of cognitive processes, each having a dual information source, we now characterize them in terms of equivalence classes leading to groupoid structures. Subsequent sections examine dynamical and phase transition behavior in these and other 'language' systems with associated information sources.

2.1 The cognitive modular network symmetry groupoid

A formal equivalence class algebra can be constructed for a cognitive process characterized by a dual information source by choosing different origin points a_0, in the sense of section 1.2 above, and defining equivalence of two states by the existence of a high-probability meaningful path connecting them with the same origin. Disjoint partition by equivalence class, analogous to orbit equivalence classes for dynamical systems, defines the vertices of a network of cognitive dual languages. Each vertex then represents a different information source dual to a cognitive process. This is not a direct representation as in a neural network, or of some circuit in silicon. It is, rather, an abstract set of 'languages' dual to the cognitive processes instantiated by biological structures, ecosystems, social process, or their hybrids. Our particular interest, however, is in an interacting network of cognitive processes of gene expression and in their relations to embedding contexts.

This structure generates a groupoid, in the sense of Weinstein (1996). States a_j, a_k in a set A are related by the groupoid morphism if and only

R. Wallace et al., *Farming Human Pathogens*, DOI 10.1007/978-0-387-92213-3_2,
© Springer Science+Business Media, LLC 2009

if there exists a high-probability grammatical path connecting them to the same base point, and tuning across the various possible ways in which that can happen – the different cognitive languages – parametizes the set of equivalence relations and creates the groupoid. This assertion requires some development.

Note that not all possible pairs of states (a_j, a_k) can be connected by such a morphism, that is, by a high-probability, grammatical and syntactical path linking them with some given base point. Those that can define the groupoid element, a morphism $g = (a_j, a_k)$ having the natural inverse $g^{-1} = (a_k, a_j)$. Given such a pairing, it is possible to define 'natural' end-point maps $\alpha(g) = a_j, \beta(g) = a_k$ from the set of morphisms G into A, and a formally associative product in the groupoid $g_1 g_2$ provided $\alpha(g_1 g_2) = \alpha(g_1), \beta(g_1 g_2) = \beta(g_2)$, and $\beta(g_1) = \alpha(g_2)$. Then the product is defined, and associative, $(g_1 g_2)g_3 = g_1(g_2 g_3)$.

In addition, there are natural left and right identity elements λ_g, ρ_g such that $\lambda_g g = g = g \rho_g$ (Weinstein, 1996).

An orbit of the groupoid G over A is an equivalence class for the relation $a_j \sim G a_k$ if and only if there is a groupoid element g with $\alpha(g) = a_j$ and $\beta(g) = a_k$. Following Cannas da Silva and Weinstein (1999), we note that a groupoid is called transitive if it has just one orbit. The transitive groupoids are the building blocks of groupoids in that there is a natural decomposition of the base space of a general groupoid into orbits. Over each orbit there is a transitive groupoid, and the disjoint union of these transitive groupoids is the original groupoid. Conversely, the disjoint union of groupoids is itself a groupoid.

The isotropy group of $a \in X$ consists of those g in G with $\alpha(g) = a = \beta(g)$. These groups prove fundamental to classifying groupoids.

If G is any groupoid over A, the map $(\alpha, \beta) : G \to A \times A$ is a morphism from G to the pair groupoid of A. The image of (α, β) is the orbit equivalence relation $\sim G$, and the functional kernel is the union of the isotropy groups. If $f : X \to Y$ is a function, then the kernel of f, $ker(f) = [(x_1, x_2) \in X \times X : f(x_1) = f(x_2)]$ defines an equivalence relation.

Groupoids may have additional structure. As Weinstein (1996) explains, a groupoid G is a topological groupoid over a base space X if G and X are topological spaces and α, β and multiplication are continuous maps. A criticism sometimes applied to groupoid theory is that their classification up to isomorphism is nothing other than the classification of equivalence relations via the orbit equivalence relation and groups via the isotropy groups. The imposition of a compatible topological structure produces a nontrivial interaction between the two structures. Below we will introduce a metric structure on manifolds of related information sources, producing such interaction.

In essence, a groupoid is a category in which all morphisms have an inverse, here defined in terms of connection to a base point by a meaningful path of an information source dual to a cognitive process.

As Weinstein (1996) points out, the morphism (α, β) suggests another way of looking at groupoids. A groupoid over A identifies not only which elements of A are equivalent to one another (isomorphic), but *it also parametizes the different ways (isomorphisms) in which two elements can be equivalent*, i.e., all possible information sources dual to some cognitive process. Given the information theoretic characterization of cognition presented above, this produces a full modular cognitive network in a highly natural manner.

Brown (1987) describes the fundamental structure as follows:

> "A groupoid should be thought of as a group with many objects, or with many identities... A groupoid with one object is essentially just a group. So the notion of groupoid is an extension of that of groups. It gives an additional convenience, flexibility and range of applications...
>
> EXAMPLE 1. A disjoint union [of groups] $G = \cup_\lambda G_\lambda, \lambda \in \Lambda$, is a groupoid: the product ab is defined if and only if a, b belong to the same G_λ, and ab is then just the product in the group G_λ. There is an identity 1_λ for each $\lambda \in \Lambda$. The maps α, β coincide and map G_λ to λ, $\lambda \in \Lambda$.
>
> EXAMPLE 2. An equivalence relation R on [a set] X becomes a groupoid with $\alpha, \beta : R \rightarrow X$ the two projections, and product $(x, y)(y, z) = (x, z)$ whenever $(x, y), (y, z) \in R$. There is an identity, namely (x, x), for each $x \in X$..."

Weinstein (1996) makes the following fundamental point:

> "Almost every interesting equivalence relation on a space B arises in a natural way as the orbit equivalence relation of some groupoid G over B. Instead of dealing directly with the orbit space B/G as an object in the category S_{map} of sets and mappings, one should consider instead the groupoid G itself as an object in the category G_{htp} of groupoids and homotopy classes of morphisms."

Later we will explore homotopy in paths generated by information sources.

The groupoid approach has become quite popular in the study of networks of coupled dynamical systems which can be defined by differential equation models, (Golubitsky and Stewart, 2006; Stewart et al. 2003; Stewart, 2004). Here we have outlined how to extend the technique to networks of interacting information sources which, in a dual sense, characterize cognitive processes, and cannot at all be described by the usual differential equation models. These latter, it seems, are much the spiritual offspring of 18th Century mechanical clock models. Cognitive processes in biological or social systems involve neither computers nor clocks, but remain constrained by the limit theorems of information theory, and these permit scientific inference on necessary conditions.

2.2 Global and local symmetry groupoids

Here we follow Weinstein (1996) fairly closely, using his example of a finite tiling.

Consider a tiling of the euclidean plane R^2 by identical 2 by 1 rectangles, specified by the set X (one dimensional) where the grout between tiles is $X = H \cup V$, having $H = R \times Z$ and $V = 2Z \times R$, where R is the set of real numbers and Z the integers. Call each connected component of $R^2 \backslash X$, that is, the complement of the two dimensional real plane intersecting X, a tile.

Let Γ be the group of those rigid motions of R^2 which leave X invariant, i.e., the normal subgroup of translations by elements of the lattice $\Lambda = H \cap V = 2Z \times Z$ (corresponding to corner points of the tiles), together with reflections through each of the points $1/2\Lambda = Z \times 1/2Z$, and across the horizontal and vertical lines through those points. As noted by Weinstein (1996), much is lost in this coarse-graining, in particular the same symmetry group would arise if we replaced X entirely by the lattice Λ of corner points. Γ retains no information about the local structure of the tiled plane. In the case of a real tiling, restricted to the finite set $B = [0, 2m] \times [0, n]$ the symmetry group shrinks drastically: The subgroup leaving $X \cap B$ invariant contains just four elements even though a repetitive pattern is clearly visible. A two-stage groupoid approach recovers the lost structure.

We define the transformation groupoid of the action of Γ on R^2 to be the set

$$G(\Gamma, R^2) = \{(x, \gamma, y | x \in R^2, y \in R^2, \gamma \in \Gamma, x = \gamma y\},$$

with the partially defined binary operation

$$(x, \gamma, y)(y, \nu, z) = (x, \gamma\nu, z).$$

Here $\alpha(x, \gamma, y) = x$, and $\beta(x, \gamma, y) = y$, and the inverses are natural.

We can form the restriction of G to B (or any other subset of R^2) by defining

$$G(\Gamma, R^2)|_B = \{g \in G(\Gamma, R^2) | \alpha(g), \beta(g) \in B\}$$

[1]. An orbit of the groupoid G over B is an equivalence class for the relation

$x \sim_G y$ if and only if there is a groupoid element g with $\alpha(g) = x$ and $\beta(g) = y$.

Two points are in the same orbit if they are similarly placed within their tiles or within the grout pattern.

[2]. The isotropy group of $x \in B$ consists of those g in G with $\alpha(g) = x = \beta(g)$. It is trivial for every point except those in $1/2\Lambda \cap B$, for which it is $Z_2 \times Z_2$, the direct product of integers modulo two with itself.

By contrast, embedding the tiled structure within a larger context permits definition of a much richer structure, i.e., the identification of local symmetries.

We construct a second groupoid as follows. Consider the plane R^2 as being decomposed as the disjoint union of $P_1 = B \cap X$ (the grout), $P_2 = B \backslash P_1$ (the complement of P_1 in B, which is the tiles), and $P_3 = R^2 \backslash B$ (the exterior of the tiled room). Let E be the group of all euclidean motions of the plane, and define the local symmetry groupoid G_{loc} as the set of triples (x, γ, y) in $B \times E \times B$ for which $x = \gamma y$, and for which y has a neighborhood $\mathcal{U}$ in R^2 such that $\gamma(\mathcal{U} \cap P_i) \subseteq P_i$ for $i = 1, 2, 3$. The composition is given by the same formula as for $G(\Gamma, R^2)$.

For this groupoid-in-context there are only a finite number of orbits:

$\mathcal{O}_1 = $ interior points of the tiles.

$\mathcal{O}_2 = $ interior edges of the tiles.

$\mathcal{O}_3 = $ interior crossing points of the grout.

$\mathcal{O}_4 = $ exterior boundary edge points of the tile grout.

$\mathcal{O}_5 = $ boundary 'T' points.

$\mathcal{O}_6 = $ boundary corner points.

The isotropy group structure is, however, now very rich indeed:

The isotropy group of a point in $\mathcal{O}_1$ is now isomorphic to the entire rotation group O_2.

It is $Z_2 \times Z_2$ for $\mathcal{O}_2$.

For $\mathcal{O}_3$ it is the eight-element dihedral group D_4.

For $\mathcal{O}_4, \mathcal{O}_5$ and $\mathcal{O}_6$ it is simply Z_2.

These are the 'local symmetries' of the tile-in-context.

Next we will attempt to create a 'biopsychosociocultural' model for single-workspace processes using just such a nested hierarchy, which splits the simple groupoid modular network into a much more complicated structure.

2.3 Internal forces breaking the symmetry groupoid

The symmetry groupoid, as we have constructed it for cognitive modules, in a kind of information space, is parametized across that space by the possible ways in which states a_j, a_k can be equivalent, that is, connected to some origin by a meaningful path of an information source dual to a cognitive process. These are different, and in this approximation, non-interacting cognitive processes. But symmetry groupoids, like symmetry groups, are made to be broken. By internal cross-talk akin to spin-orbit interactions within a symmetric atom, and by cross-talk with slower, external, information sources, akin to putting a symmetric atom in a powerful magnetic or electric field.

As to the first process, suppose that linkages can fleetingly occur between the ordinarily disjoint cognitive modules defined by the network groupoid. In the spirit of Wallace (2005a), this is represented by establishment of a non-zero mutual information measure between them: a cross-talk which breaks the strict groupoid symmetry developed above.

Wallace (2005a) describes this structure in terms of fixed magnitude disjunctive strong ties which give the equivalence class partitioning of modules, and nondisjunctive weak ties which link modules across the partition, and parametizes the overall structure by the average strength of the weak ties, to use Granovetter's (1973) term. By contrast the approach of Wallace (2005b), which we outline here, is to simply look at the average number of fixed-strength nondisjunctive links in a random topology. These are obviously two analytically tractable limits of a much more complicated regime of possible models.

Since we know nothing about how the cross-talk connections can occur, we will – at first – assume they are random and construct a random graph in the classic Erdos-Renyi manner.

Suppose there are M disjoint cognitive modules – M elements of the equivalence class algebra of languages dual to some cognitive process – which we now take to be the vertices of a possible graph. For M very large, following Savante et al. (1993), when edges (defined by establishment of a fixed-strength mutual information measure between the graph vertices) are added at random to M initially disconnected vertices, a remarkable transition occurs when the number of edges becomes approximately $M/2$. Erdos and Renyi (1960) studied random graphs with M vertices and $(M/2)(1 + \mu)$ edges as $M \rightarrow \infty$, and discovered that such graphs almost surely have the following properties (Molloy and Reed, 1995, 1998; Grimmett and Stacey, 1998; Luczak, 1990; Aiello et al., 200; Albert and Barabasi, 2002):

[1] If $\mu < 0$, only small trees and unicyclic components are present, where a unicyclic component is a tree with one additional edge. Moreover, the size of the largest tree component is $(\mu - \ln(1 + \mu))^{-1} + \mathcal{O}(\log \log n)$.

[2] If $\mu = 0$, however, the largest component has size of order $M^{2/3}$.

[3] If $\mu > 0$, there is a unique giant component (GC) whose size is of order M; in fact, the size of this component is asymptotically αM, where $\mu = -\alpha^{-1}[\ln(1 - \alpha) - 1]$, which has an explicit solution for α in terms of the Lambert W-function. Thus, for example, a random graph with approximately $M \ln(2)$ edges will have a giant component containing $\approx M/2$ vertices.

Such a phase transition initiates a new, collective, cognitive phenomenon. At the level of the individual mind, as opposed to the coupled networks of gene expression which concern us here, unconscious cognitive modules link to become tunable general broadcasts, emergently defined by a set of cross-talk mutual information measures between interacting, lower level, cognitive submodules. The source uncertainty, H, of the language dual to the collective cognitive process, which characterizes the richness of the cognitive language of the workspace, will grow as some monotonic function of the size of the GC, as more and more unconscious processes are incorporated into it. Wallace (2005b) provides details.

Others have taken similar network phase transition approaches to assemblies of neurons. There is, for example the neuropercolation work of Kozma and colleagues (Kozma et al., 2004, 2005), but their approach has not focused explicitly on modular networks of cognitive processes, which may or may not be instantiated by neurons. Restricting analysis to such modular networks finesses much of the underlying conceptual difficulty, and permits use of the asymptotic limit theorems of information theory and the import of techniques from statistical physics, a matter we will discuss later.

2.4 External forces breaking the symmetry groupoid

Just as a higher order information source, associated with the giant component of a random or semirandom graph, can be constructed out of the interlinking of cognitive modules by mutual information, so too external information sources, for example other physiological systems, and/or embedding ecosystem or sociocultural structures, can be represented as slower-acting information sources whose influence on the GC can be felt in a collective mutual information measure. These constitute an onion-like 'structured environment', analogous to Baars' contexts affecting individual consciousness (Baars, 1988, 2005; Baars and Franklin, 2003). The collective mutual information measure will, through the Joint Asymptotic Equipartition Theorem, which generalizes the Shannon-McMillan Theorem, be the splitting criterion for high and low probability joint paths across the entire system.

The tool for this is network information theory (Cover and Thomas, 1991, p. 388). Given three interacting information sources, Y_1, Y_2, Z, the splitting criterion, taking Z as the 'external context', is given by

$$I(Y_1, Y_2|Z) = H(Z) + H(Y_1|Z) + H(Y_2|Z) - H(Y_1, Y_2, Z),$$

(2.1)

where $H(..|..)$ and $H(..,..,..)$ represent conditional and joint uncertainties (Khinchin, 1957; Ash, 1990; Cover and Thomas, 1991).

This generalizes to

$$I(Y_1, ...Y_n|Z) = H(Z) + \sum_{j=1}^{n} H(Y_j|Z) - H(Y_1, ..., Y_n, Z).$$

(2.2)

If we assume the general giant component of gene expression involves a very rapidly shifting, and perhaps even highly tunable, dual information source X, embedding contextual factors will have a set of significantly slower-responding sources $Y_j, j = 1..m$, and external social, cultural, and other environmental processes will be characterized by even more slowly-acting sources $Z_k, k = 1..n$. Mathematical induction on equation (2.2) gives a complicated expression for a mutual information splitting criterion which we write as

$$I(X|Y_1, .., Y_m|Z_1, .., Z_n).$$

(2.3)

Presumably this could be extended in a natural manner to incorporate a set of rapid information sources $X_1, X_2, ...$, giving something having the form

$$I(X_1, .., X_i|Y_1, .., Y_m|Z_1, .., Z_n).$$

Equation (2.3) and its generalizations encompass a fully interpenetrating structure for gene expression and other cognitive phenomena, one in which analogs to Baars' contexts act as important, but flexible, boundary conditions, defining the underlying topology available to a far more rapidly shifting analog to the global workspace of high order mental function.

This result does not commit the mereological fallacy (Bennett and Hacker, 2003), that is, the mistake of imputing to a part of a system the characteristics which require functional entirety. Epigenetics and phenotype formation arise from more than individual gene activation.

2.5 Emergence in information systems as a phase transition

As a number of researchers have noted, in one way or another, – see Wallace, (2005a) or Wallace and Fullilove, (2008) for discussion – equation (1.5),

$$H \equiv \lim_{n \to \infty} \frac{\log[N(n)]}{n},$$

is homologous to the thermodynamic limit in the definition of the free energy density of a physical system. This has the form

$$F(K) = \lim_{V \to \infty} \frac{\log[Z(K)]}{V},$$

(2.4)

where F is the free energy density, K the inverse temperature, V the system volume, and $Z(K)$ is the partition function defined by the system Hamiltonian. Any good statistical mechanics text will provide details (e.g., Landau and Lifshitz, 2007).

The next chapter shows at some length how this homology permits the natural transfer of standard renormalization methods from statistical mechanics to information theory, producing phase transitions and analogs to evolutionary punctuation in systems characterized by adiabatically, piecewise stationary, ergodic information sources. Such biological phase changes appear to be ubiquitous. Wallace (2002) uses these arguments to explore the differences and similarities between evolutionary punctuation in genetic and learning plateaus in neural systems.

The approach uses a mean field approximation in which average strength (or probability) of nondisjunctive linkages between cognitive nodes – crosstalk – serves as a kind of inverse temperature parameter. Phase transitions can then be described using various 'biological' renormalization strategies, in which universality class tuning becomes the principal second order mechanism. These are continuous phase transitions in that there is no latent heat required, as in boiling water to steam, although the treatment of resilience in cognitive dynamic manifolds which will emerge in due course can perhaps be viewed as an analog. Under a resilience domain shift the underlying system topology changes, either within or between dynamic manifolds, and this can be viewed as similar to the fundamental geometric structural change water undergoes when it transforms from crystal to liquid, or from liquid to gas. Indeed, recent work (Franzosi and Pettini, 2004; Pettini, 2007) uses Morse theory from differential topology to identify necessary conditions for topological shifts on dynamic manifolds of systems undergoing first and second order phase transition. Sufficiency, on the other hand, is an open question.

We have shown above that there is another analytically tractable limit, the giant component, suggesting the possibility of intermediate cases. In the next section we extend the Giant Component paradigm, so that the mean number of such linkages, above some variable threshold, is the parameter of central interest, and the second order tuning involves topological mechanisms similar to the approach of Morse theory.

Whatever scheme is chosen, however, the homology between equations (1.5) and (2.4) ensures that some form of emergent behavior, akin to a physical phase transition, is inevitable for networks of interacting cognitive systems, however instantiated.

2.6 Multiple workspaces: topological tuning

The random network development above is predicated on there being a variable average number of fixed-strength linkages between components. Clearly, the mutual information measure of cross-talk is not inherently fixed, but can continuously vary in magnitude. This we address by a parametized renormalization. In essence, the modular network structure linked by mutual information interactions has a topology depending on the degree of interaction of interest. Suppose we define an interaction parameter ω, a real positive number, and look at geometric structures defined in terms of linkages which are zero if mutual information is less than, and 'renormalized' to unity if greater than, ω. Any given ω will define a regime of giant components of network elements linked by mutual information greater than or equal to it.

The fundamental conceptual trick at this point is to invert the argument. A given topology for the giant component will, in turn, define some value of ω_C, so that network elements interacting by mutual information less than that value will be unable to participate. That is, they will be locked out and not be perceived. We hence are assuming that the ω is a tunable, syntactically dependent detection limit that depends on the instantaneous topology of the giant component defining the global workspace of interest. That topology is the basic tunable syntactic filter across the underlying modular symmetry groupoid, and variation in ω is only one aspect of a much more general topological shift. More detailed analysis is given below in terms of a topological rate distortion manifold.

This argument defines level sets of ω in the sense of Morse theory. See the Mathematical Appendix for details. If we agree to parametize the driving network structures by some set of characteristic numbers, for example those of the probability distribution of network links per node, $P(k)$, say real numbers $\beta_1, ..., \beta_m$, then the topology of the m-dimensional manifold defined by that set parameters can be fully characterized by the critical points of ω. That is, the critical points of ω, those for which $d_\beta\omega = 0$, can be used not only to determine when there is a fundamental shift in 'high level cognitive topology', but to infer a good deal of its structure as well. It seems likely, however, that some extension of Morse theory to groupoid topologies will be necessary.

There is considerable empirical evidence from fMRI brain imaging experiments to show that individual human consciousness involves a single, shifting, global general broadcast, a matter leading necessarily to the phenomenon of inattentional blindness. Cognitive submodules within institutions – individuals, departments, formal and informal workgroups – by contrast, can do more than one thing, and indeed, are usually required to multitask. The intent of this work is to suggest the possibility that interacting networks of gene expression may behave in a recognizably similar manner, acting under similar necessary conditions constraints imposed by the asymptotic limit theorems of information theory.

Evidently multiple workspaces would lessen the probability of some analog to inattentional blindness (Wallace, 2007), but, we will find, do not eliminate it. Indeed, these introduce other failure modes, in particular failure of communication between multiple giant components of gene expression, leading to 'Rate Distortion' errors. In a later section we will examine the way in which inattentional blindness and rate distortion error might interact to create pathologies of gene expression under conditions of resource limitation and/or externally-imposed stress.

We must postulate a set of crosstalk information measures between cognitive submodules, each associated with its own tunable giant component having its own special topology.

Suppose the set of giant components at some 'time' k is characterized by a set of parameters $\Omega_k \equiv \omega_1^k, ..., \omega_m^k$. Fixed parameter values define a particular giant component set having a particular topological structure, in the Morse theory sense. Suppose that over a sequence of 'times' the set of giant components can be characterized by a (possibly coarse-grained) path $x_n = \Omega_0, \Omega_1, ..., \Omega_{n-1}$ having significant serial correlations. This permits us to define an adiabatically, piecewise stationary, ergodic (APSE) information source. Call that information source $\mathbf{X}$.

Suppose that a set of (external or internal) signals impinging on the set of giant components is also highly structured and forms another APSE information source $\mathbf{Y}$ which interacts not only with the system of interest globally, but specifically with the tuning parameters of the set of giant components characterized by $\mathbf{X}$. $\mathbf{Y}$ is necessarily associated with a set of paths y_n.

Pair the two sets of paths into a joint path $z_n \equiv (x_n, y_n)$, and invoke some inverse coupling parameter, K, between the information sources and their paths. By arguments summarized in the next chapter, this leads to phase transition punctuation of $I[K]$, the mutual information between $\mathbf{X}$ and $\mathbf{Y}$, under either the Joint Asymptotic Equipartition Theorem, or, given a distortion measure, under the Rate Distortion Theorem.

$I[K]$ is a splitting criterion between high and low probability pairs of paths, and partakes of the homology with free energy density. Attentional focusing then itself becomes a punctuated event in response to increasing linkage between the structure of interest and an external signal, or some particular system of internal events. This iterated argument parallels the extension of the General Linear Model into the Hierarchical Linear Model of regression theory.

Call this a Multitasking Hierarchical Cognitive Model (MHCM). For a simple organism, there may be only one giant component. Those more complex may have a large, and even very large, set of them.

As briefly described, this requirement leads to new potential failure modes related to impaired communication between giant components.

That is, a complication specific to high order cognitive phenomena lies in the necessity of information transfer between giant components. The form and function of such interactions will, of course, be determined by the nature of the particular system, but, synchronous or asynchronous, contact between giant components is circumscribed by the Rate Distortion Theorem. That theorem, reviewed in the Mathematical Appendix, states that, for a given maximum acceptable critical average signal distortion, there is a limiting minimum information transmission rate – channel capacity – such that messages sent at rates greater than that limit are guaranteed to have average distortion less than the critical maximum. Too rapid transmission between parallel global workspaces having limited channel capacity linking them – information overload – violates that condition, and guarantees large signal distortion. This is a likely failure mode that appears unique to multiple workspace systems. We will argue these may otherwise have a lessened probability of inattentional blindness, of overfocus analogous to the 'no free lunch' conundrum of optimization theory, described in a later section.

2.7 Phenomenological Landau theory

The homology between equations 1.5 and 2.4 suggests the possibility of abducting standard techniques from statistical physics into the analysis of cognitive process, however instantiated. Here we closely follow the development in Skierski et al. (1989, p. 3789).

Most simply Landau's theory of phase transitions (Landau and Lifshitz, 2007) assumes that the free energy of a system near criticality can be expanded in a power series of some 'order parameter' ϕ which represents a fundamental measurable quantity, that is, a symmetry invariant. One writes

$$F_0 = \sum_{k=m}^{p(>m)} A_k \phi^k,$$

(2.5)

with $A_2 \approx \alpha(T - T_c)$ sufficiently close to the critical temperature T_c. This mean field approach can be used to describe a variety of second-order effects for $p = 4$ or $p = 6$, $A_3 = 0$ and $A_4 > 0$, and first order phase transitions (requiring latent heat) for either $p = 6, A_3 = 0, A_4 < 0$ or $p = 4$ and $A_3 \neq 0$. These can be both temperature induced (for $m = 2$) and field induced (for $m = 1$).

Minimization of F_0 with respect to the order parameter yields the average value of ϕ, $< \phi >$, which is zero above the critical temperature and non-zero below it. In the absence of external fields, the second-order transition occurs at $T = T_c$, while the first-order, needing latent heat, occurs at $T_c^* = T_c + A_4^2/4\alpha A_6$. In the latter case thermal hysteresis arises between $T_s \equiv T_c + A_4^2/3\alpha A_6$ and T_c. A more accurate approximation involves an expression that recognizes the effect of coarse-graining, adding a term in $\nabla^2 \phi$ and integrating over space rather than summing. Regimes dominated by this gradient will show behaviors analogous to those described using the one dimensional Landau-Ginzburg equation, which, among other things, characterizes superconductivity.

The analogy between free energy density and information source uncertainty – replacing integration over volume by the sum over n – suggests examining the dynamics of some empirical, quantitative 'order parameter' characterizing large-scale cognitive function near a threshold driven by an average level of crosstalk between cognitive modules. Such behavior can be expressed in terms of an equation similar to that above, having the form

$$H \approx \sum_{k=m}^{p(>m)} A_k \phi^k.$$

(2.6)

ϕ would then be some index of a system's collective cognition. T is then an average inverse crosstalk measure.

Note that it is possible to apply a Morse theory approach at this juncture (Michel and Mozrzymas, 1977).

The Landau formalism quickly enters deep topological waters (Pettini, 2007, pp. 42-43; Landau and Lifshitz, 2007, pp. 459-466). The essence of Landau's insight was that phase transitions without latent heat – second order transitions – were usually in the context of a significant symmetry change in the physical states of a system, with one phase being far more symmetric than the other. A symmetry is lost in the transition, a phenomenon called spontaneous symmetry breaking. The greatest possible set of symmetries in a physical system is that of the Hamiltonian describing its energy states. Usually states accessible at lower temperatures will lack symmetries available at higher temperatures, so that the lower temperature phase is the less symmetric: The randomization of higher temperatures ensures that higher symmetry/energy states will then be accessible to the system.

At the lower temperature an order parameter must be introduced to describe the system's physical states – some extensive quantity like magnetization. The order parameter will vanish at higher temperatures, involving more symmetric states, and will be different from zero in the less symmetric lower temperature phase.

This can be formalized, following Pettini (2007), as follows. Consider a thermodynamic system having a free energy F which is a function of temperature T, pressure P, and some other extensive macroscopic parameters m_i, so that $F = F(P, T, m_i)$. The m_i all vanish in the most symmetric phase, so that, as a function of the m_i, $F(P, T, m_i)$ is invariant with respect to the transformations of the symmetry group G_0 of the most symmetric phase of the system when all $m_i \equiv 0$.

The state of the system can be represented by a vector $|m>=|m_1,...,m_n>$ in a vector space $\mathcal{E}$. Now, within $\mathcal{E}$, construct a linear representation of the group G_0 that associates with any $g \in G_0$ a matrix $M(g)$ having rank n. In general, the representation $M(g)$ is reducible, and we can decompose $\mathcal{E}$ into invariant irreducible subspaces $\mathcal{E}_1, \mathcal{E}_2, ..., \mathcal{E}_k$, having basis vectors $|e_i^{(n)}>$ with $n = 1, 2, ...n_i$ and $n_i = dim\mathcal{E}_i$. The state variables m_i are transformed into new variables $\eta_i^{(n)} = <e_i^{(n)}|m>$, where the bracket represents an inner product.

In terms of irreducible representations $D_i(g)$ induced by $M(g)$ in $\mathcal{E}_i$ we have

$$M(g) = D_1(g) \oplus D_2(g)\oplus, ..., \oplus D_k(g).$$

If at least one of the $\eta_i^{(n)}$ is nonzero, then the system no longer has the symmetry G_0. This symmetry has been broken, and the new symmetry group is G_i, associated with the representation $D_i(g)$ in $\mathcal{E}_i$. The variables $\eta_i^{(n)}$ are the new order parameters, and the free energy is now $F = F(P, T, \eta_i^{(n)})$. For a physical system the actual values of the η as functions of P and T can be variationally determined by minimizing the free energy F.

Two essential features distinguish cognitive systems from this simple physical model.

First, order parameters cannot be determined by simple minimization procedures, as cognitive systems can, within their contextual constraints (which include available energy), choose states which are not energy or other extrema.

Second, the essential symmetry of information sources dual to cognitive process is driven by groupoid, rather than group, structures. One must then engage the full transitive orbit/isotropy group decomposition, and examine groupoid representations which are configured about the irreducible representations of the isotropy groups. We will not explore this complication further.

More fundamentally, however, Landau's theory is not an 'edge-of-chaos' model in the sense that it is a mean field analysis neglecting the thermodynamic fluctuations that become significant near phase transition. Wilson's (1971) adaptation of renormalization formalism to phase transitions remedies the omission. Indeed the remedy provides the basis for the extended approach of the following chapter where, not only are renormalization symmetries themselves changeable, but within a single symmetry, universality class tuning becomes possible, allowing the tuning of punctuated accession to higher order cognitive function in a second order model.

2.8 The dynamical groupoid: Phenomenological Onsager theory

A fundamental homology between the information source uncertainty dual to a cognitive process and the free energy density of a physical system arises, in part, from the formal similarity between their definitions in the asymptotic limit. Information source uncertainty can be defined as in equation (1.5). This is, as noted, quite analogous to the free energy density of a physical system, equation (2.4).

Feynman (1996) provides a series of physical examples, based on Bennett's work, where this homology is, in fact, an identity, at least for very simple systems. Bennett argues, in terms of idealized irreducibly elementary computing machines, that the information contained in a message can be viewed as the work saved by not needing to recompute what has been transmitted.

Feynman explores in some detail Bennett's ideal microscopic machine designed to extract useful work from a transmitted message. The essential argument is that computing, in any form, takes work. Thus the more complicated a cognitive process, measured by its information source uncertainty, the greater its energy consumption, and our ability to provide energy to any cognitive process is limited. For example, a unit of brain tissue consumes an order of magnitude more energy than a unit of any other tissue. Inattentional blindness, Wallace (2007) argues, emerges as a thermodynamic limit on processing capacity in a topologically-fixed global workspace, that is, one which has been strongly configured about a particular task. Generalizations seem straightforward.

Understanding the time dynamics of cognitive systems away from the kind of phase transition critical points described in section 2.5 requires a phenomenology similar to the Onsager relations of nonequilibrium thermodynamics. This will lead to a more general phase transition theory involving another set of large-scale topological changes in the sense of Morse theory.

If the dual source uncertainty of a cognitive process is parametized by some vector of quantities $\mathbf{K} \equiv (K_1, ..., K_m)$, then, in analogy with nonequilibrium thermodynamics, gradients in the K_j of the *disorder*, defined as

$$S \equiv H(\mathbf{K}) - \sum_{j=1}^{m} K_j \partial H / \partial K_j$$

(2.7)

become of central interest.

Equation (2.7) is similar to the definition of entropy in terms of the free energy density of a physical system, as suggested by the homology between free energy density and information source uncertainty described above.

Pursuing the homology further, the generalized Onsager relations defining temporal dynamics become

$$dK_j/dt = \sum_i L_{j,i} \partial S / \partial K_i,$$

(2.8)

where the $L_{j,i}$ are, in first order, constants reflecting the nature of the underlying cognitive phenomena. The L-matrix is to be viewed empirically, in the same spirit as the slope and intercept of a regression model, and may have structure far different than more familiar simple chemical or physical processes. The $\partial S / \partial K$ are analogous to thermodynamic forces in a chemical system, and may be subject to override by external physiological or ecological driving mechanisms, a matter pursued further in section 2.14 below.

An essential contrast with simple physical systems driven by, say, entropy maximization is that cognitive systems make decisions about resource allocation, to the extent resources are available. That is, resource availability is a *context* for cognitive function, in the sense of Baars, not a *determinant*.

Equations (2.7) and (2.8) can be derived in a simple parameter-free covariant manner which relies on the underlying topology of the information source space implicit to the development. Different cognitive phenomena have, according to our development, dual information sources, and we are interested in the local properties of the system near a particular reference state. We impose a topology on the system, so that, near a particular 'language' A, dual to an underlying cognitive process, there is (in some sense) an open set U of closely similar languages $\hat{A}$, such that $A, \hat{A} \subset U$. Note that it may be necessary to coarse-grain the system's responses to define these information sources. The problem is to proceed in such a way as to preserve the underlying essential topology, while eliminating 'high frequency noise'. The formal tools for this can be found elsewhere, e.g., in Chapter 8 of Burago et al. (2001).

Since the information sources dual to the cognitive processes are similar, for all pairs of languages $A, \hat{A}$ in U, it is possible to:

[1] Create an embedding alphabet which includes all symbols allowed to both of them.

[2] Define an information-theoretic distortion measure in that extended, joint alphabet between any high probability (grammatical and syntactical) paths in A and $\hat{A}$, which we write as $d(Ax, \hat{A}x)$ (Cover and Thomas, 1991, and the appendix). Note that these languages do not interact, in this approximation.

[3] Define a metric on U, for example,

$$\mathcal{M}(A, \hat{A}) = |\lim \frac{\int_{A,\hat{A}} d(Ax, \hat{A}x)}{\int_{A,A} d(Ax, A\hat{x})} - 1|,$$

(2.9)

using an appropriate integration limit argument over the high probability paths. Note that the integration in the denominator is over different paths within A itself, while in the numerator it is between different paths in A and $\hat{A}$.

Consideration suggests $\mathcal{M}$ is a formal metric, having

$$\mathcal{M}(A, B) \geq 0, \mathcal{M}(A, A) = 0, \mathcal{M}(A, B) = \mathcal{M}(B, A),$$

$$\mathcal{M}(A, C) \leq \mathcal{M}(A, B) + \mathcal{M}(B, C).$$

Other approaches to metric construction on U seem possible.

Structures weaker than a conventional metric would be of more general utility, but the mathematical complications are formidable (Glazebrook and Wallace, 2007).

Note that these conditions can be used to define equivalence classes of *languages*, where previously we defined equivalence classes of *states* which could be linked by high probability, grammatical and syntactical paths to some base point. This led to the characterization of different information sources. Here we construct an entity, formally a topological manifold, *which is an equivalence class of information sources*. This is, provided $\mathcal{M}$ is a conventional metric, a classic differentiable manifold. The set of such equivalence classes generates the *dynamical groupoid*, and questions arise regarding mechanisms, internal or external, which can break that groupoid symmetry, as in the previous example. In particular, imposition of a metric structure on this groupoid, and on its base set, would permit a nontrivial interaction between orbit equivalence relations and isotropy groups, leading to interesting algebraic structures.

Since H and $\mathcal{M}$ are both scalars, a 'covariant' derivative can be defined directly as

$$dH/d\mathcal{M} = \lim_{\hat{A} \to A} \frac{H(A) - H(\hat{A})}{\mathcal{M}(A, \hat{A})},$$

(2.10)

where $H(A)$ is the source uncertainty of language A.

Suppose the system to be set in some reference configuration A_0.

To obtain the unperturbed dynamics of that state, impose a Legendre transform using this derivative, defining another scalar

$$S \equiv H - \mathcal{M} dH/d\mathcal{M}.$$

(2.11)

The simplest possible Onsager relation – here seen as an empirical, fitted, equation like a regression model – in this case becomes

$$d\mathcal{M}/dt = L dS/d\mathcal{M},$$

(2.12)

where t is the time and $dS/d\mathcal{M}$ represents an analog to the thermodynamic force in a chemical system. This is seen as acting on the reference state A_0. For

$$dS/d\mathcal{M}|_{A_0} = 0,$$

$$d^2 S/d\mathcal{M}^2|_{A_0} > 0,$$

(2.13)

the system is quasistable, a Black hole, if you will, and externally imposed forcing mechanisms will be needed to effect a transition to a different state. We shall explore this circumstance below in terms of topological considerations analogous to the concept of ecosystem resilience.

Conversely, changing the direction of the second condition, so that

$$dS^2/d\mathcal{M}^2|_{A_0} < 0,$$

leads to a repulsive peak, a White hole, representing a possibly unattainable realm of states.

Explicit parametization of $\mathcal{M}$ introduces standard – and quite considerable – notational complications (Burago et al., 2001; Auslander, 1967): Imposing a metric for different cognitive dual languages parametized by $\mathbf{K}$ leads to Riemannian, or even Finsler, geometries, including the usual geodesics. See the Mathematical Appendix for details.

The dynamics, as we have presented them so far, have been noiseless, while neural systems, from which we are abducting theory, are well known to be very noisy, and indeed may be subject to mechanisms of stochastic resonance. Equation (2.13) might be rewritten as

$$d\mathcal{M}/dt = LdS/d\mathcal{M} + \sigma W(t),$$

where σ is a constant and $W(t)$ represents white noise. Again, S is seen as a function of the parameter $\mathcal{M}$. This leads directly to a family of classic stochastic differential equations having the form

$$d\mathcal{M}_t = L(t, \mathcal{M})dt + \sigma(t, \mathcal{M})dB_t,$$

(2.14)

where L and σ are appropriately regular functions of t and $\mathcal{M}$, and dB_t represents the noise structure, characterized by its quadratic variation. See the Mathematical Appendix for details.

In the sense of Emery (1989), this leads into deep realms of stochastic differential geometry and related topics. The obvious inference is that noise, which need not be 'white', can serve as a tool to shift the system between various equivalence classes, in essence as a kind of crosstalk and the source of a generalized stochastic resonance. See the second mathematical appendix for a brief review of stochastic differential equations.

We have defined a groupoid for the system based on a particular set of equivalence classes of information sources dual to cognitive processes. That groupoid parsimoniously characterizes the available dynamical manifolds, and, in precisely the sense of the earlier development, breaking of the groupoid symmetry creates more complex objects of considerable interest, which will be studied below. This leads to the possibility, indeed, the necessity of *Deus ex Machina* mechanisms – analogous to programming, stochastic resonance, etc. – to force transitions between the different possible modes within and across dynamic manifolds. See section 2.14. In one model the external 'programmer' creates the manifold structure, and the system hunts within that structure for the 'solution' to the problem according to equivalence classes of paths on the manifold. Noise, as with random mutation in evolutionary algorithms, might well be needed to avoid unstable equilibria (but not analogs to limit cycles or pseudorandom strange attractors and their generalizations).

Equivalence classes of *states* gave dual information sources. Equivalence classes of *information sources* give different characteristic system dynamics, representing different 'programs'. Later we will examine equivalence classes of *paths*, which will produce different directed homotopy topologies characterizing those dynamical manifolds. This introduces the possibility of having different quasi-stable resilience modes *within* individual dynamic manifolds. One set of these can be characterized as leading to solutions of the underlying (broadly defined) 'computing problem', while others may simply be pathological absorbing states. Pink or white noise might provide a tunable means of creating crosstalk between different topological states within a dynamical manifold, or between different dynamical manifolds altogether.

Effectively, topological shifts between and within dynamic manifolds constitute a theory of phase transitions in cognitive systems, as we have characterized them. Indeed, similar considerations have become central in the study of phase changes for physical systems. Franzosi and Pettini (2004), for instance, write,

> "The standard way of [studying phase transition in physical systems] is to consider how the values of thermodynamic observables, obtained in laboratory experiments, vary with temperature, volume, or an external field, and then to associate the experimentally observed discontinuities at a PT [phase transition] to the appearance of some kind of singularity entailing a loss of analyticity... However, we can wonder whether this is the ultimate level of mathematical understanding of PT phenomena, or if some reduction to a more basic level is possible... [Our] new theorem says that nonanalyticity is the 'shadow' of a more fundamental phenomenon occurring in configuration space: *a topology change*... [Our] theorem means that a topology change [in a particular energy manifold] is a *necessary* condition for a phase transition to take place... The topology changes implied here are those described within the framework of Morse theory through *attachment handles...*
>
> The converse of our Theorem is not true. There is not a one-to-one correspondence between phase transitions and topology changes... an open problem is that of *sufficiency* conditions, that is to determine which kinds of topology changes can entail the appearance of a [phase transition]."

The phenomenological Onsager treatment would also be enriched by adoption of a Morse theory perspective on topological transitions, following Michel and Mozrzymas (1977).

The next section introduces further canonical topological complications.

2.9 Tuning the network of dynamic manifolds

Equivalence classes of states defined a groupoid of languages dual to cognitive processes which, when broken by internal crosstalk, led to a giant component within a network of cognitive modules. Here we have introduced a new groupoid of equivalence classes of languages defining dynamical manifolds. Within a large enough system this groupoid can also be broken by crosstalk between manifolds to create another, larger, giant component, this time of linked dynamic manifolds rather than of dual languages. The giant component(s) of cognitive modules created an analog to individual consciousness which, in an institutional setting, we have elsewhere characterized as a collective consciousness (Wallace and Fullilove, 2008). This larger structure of interacting dynamic manifolds might well be designated as a higher collective cognition (HCC). Again we can can apply a Morse theory argument to a 'renormalization' cutoff of crosstalk between interacting dynamic manifolds, say $\Theta \geq 0$, which in this model becomes a Morse function, parametized by some vector of real numbers $\Gamma = (\gamma_1, ..., \gamma_s)$. Critical points of Θ, where $d_{\Gamma_c}\Theta = 0$, through the Hessian matrix in local coordinates, $((\partial^2\Theta/\partial\gamma_i\partial\gamma_j))$ at the critical points, will again parsimoniously index the underlying topology of this greater structure.

The advantage of introducing such a complication is that an HCC allows a system to radically shift gears when confronted by markedly changing environmental or other embedding patterns. The tunable giant component of available dynamical manifolds becomes a larger coping resource. Populations with a good repertoire of dynamic response surfaces would seem to fare better under regimens of shifting selection and command lower probabilities for extirpation.

Jerne's 'idiotypic network' (Cohen, 2000) of infinite mirroring between antigenic challenge and immune response appears to loom here: how many times can we iterate this structure? In point of fact, the next section shows that a system may not need much iteration at all if it can properly focus its attention.

There is a further important observation. It is clearly quite possible to introduce a multiplicity structure, leading to a family of interacting dynamic groupoids, opening the way for a vast spectrum of possible response modalities.

2.10 The rate distortion manifold

The second order iterations above, analogous to expanding the General Linear Model to the Hierarchical Linear Model, and involving paths in parameter space, can be significantly extended. This produces a generalized tunable retina model which can be interpreted as a 'Rate Distortion manifold', a concept which further opens the way for import of a vast array of tools from geometry and topology, and will lead to a parsimonious description of the famous Baldwin effect and its many possible weaker generalizations.

Suppose, now, that threshold behavior for system reaction requires some elaborate structure of nonlinear relationships defining a set of renormalization parameters $\Omega_k \equiv \omega_1^k, ..., \omega_m^k$. The critical assumption is that there is a tunable zero order state, and that changes about that state are, in first order, relatively small, although their effects on punctuated process may not be at all small. Thus, given an initial m-dimensional vector Ω_k, the parameter vector at time $k+1$, Ω_{k+1}, can, in first order, be written as

$$\Omega_{k+1} \approx \mathbf{R}_{k+1} \Omega_k,$$

(2.15)

where $\mathbf{R}_{t+1}$ is an $m \times m$ matrix, having m^2 components.

If the initial parameter vector at time $k = 0$ is Ω_0, then at time k

$$\Omega_k = \mathbf{R}_k \mathbf{R}_{k-1} ... \mathbf{R}_1 \Omega_0.$$

(2.16)

The interesting correlates of high-level cognitive function are, in this development, *now represented by an information-theoretic path defined by the sequence of operators* $\mathbf{R}_k$, each member having m^2 components. The grammar and syntax of the path defined by these operators is associated with a dual information source, in the usual manner.

The effect of an information source of external signals, $\mathbf{Y}$, is now seen in terms of more complex joint paths in Y and R-space whose behavior is, again, governed by a mutual information splitting criterion according to the JAEPT.

The complex sequence in m^2-dimensional R-space has, by this construction, been projected down onto a parallel path, the smaller set of m-dimensional ω-parameter vectors $\Omega_0, ..., \Omega_k$.

If the punctuated tuning of system attention is now characterized by a 'higher' dual information source – an embedding generalized language – so that the paths of the operators $\mathbf{R}_k$ are autocorrelated, then the autocorrelated paths in Ω_k represent output of a parallel information source. This source, given Rate Distortion limitations, is apparently a grossly simplified, and hence highly distorted, picture of the 'higher' process represented by the R-operators, having m as opposed to $m \times m$ components.

High levels of distortion may not necessarily be the case for such a structure, *provided it is properly tuned to the incoming signal.* If it is inappropriately tuned, however, then distortion may be extraordinary.

Let us examine a single iteration in more detail, assuming now there is a (tunable) zero reference state, $\mathbf{R}_0$, for the sequence of operators $\mathbf{R}_k$, and that

$$\Omega_{k+1} = (\mathbf{R}_0 + \delta\mathbf{R}_{k+1})\Omega_k,$$

(2.17)

where $\delta\mathbf{R}_k$ is 'small' in some sense compared to $\mathbf{R}_0$.

Note that in this analysis the operators $\mathbf{R}_k$ are, implicitly, determined by linear regression. We thus can invoke a quasi-diagonalization in terms of $\mathbf{R}_0$. Let $\mathbf{Q}$ be the matrix of eigenvectors which Jordan-block-diagonalizes $\mathbf{R}_0$. Then

$$\mathbf{Q}\Omega_{k+1} = (\mathbf{Q}\mathbf{R}_0\mathbf{Q}^{-1} + \mathbf{Q}\delta\mathbf{R}_{k+1}\mathbf{Q}^{-1})\mathbf{Q}\Omega_k.$$

(2.18)

If $\mathbf{Q}\Omega_k$ is an eigenvector of $\mathbf{R}_0$, say Y_j with eigenvalue λ_j, it is possible to rewrite this equation as a generalized spectral expansion

$$Y_{k+1} = (\mathbf{J} + \delta\mathbf{J}_{k+1})Y_j \equiv \lambda_j Y_j + \delta Y_{k+1}$$

$$= \lambda_j Y_j + \sum_{i=1}^{n} a_i Y_i.$$

(2.19)

$\mathbf{J}$ is a block-diagonal matrix, $\delta\mathbf{J}_{k+1} \equiv \mathbf{Q}\mathbf{R}_{k+1}\mathbf{Q}^{-1}$, and δY_{k+1} *has been expanded in terms of a spectrum of the eigenvectors of* $\mathbf{R}_0$, with

$$|a_i| \ll |\lambda_j|, |a_{i+1}| \ll |a_i|.$$

(2.20)

The point is that provided $\mathbf{R}_0$ has been tuned so that this condition is true, the first few terms in the spectrum of this iteration of the eigenstate will contain most of the essential information about $\delta\mathbf{R}_{k+1}$. This appears quite similar to the detection of color in the retina, where three overlapping non-orthogonal eigenmodes of response are sufficient to characterize a huge plethora of color sensation. Here, if such a tuned spectral expansion is possible, a very small number of observed eigenmodes would suffice to permit identification of a vast range of changes, so that the rate-distortion constraints become quite modest. That is, there will not be much distortion in the reduction from paths in R-space to paths in Ω-space. Inappropriate tuning, however, can produce very marked distortion, even inattentional blindness, in spite of multitasking.

Higher-order Rate Distortion Manifolds are likely to give better approximations than lower ones, in the same sense that second order tangent structures give better, if more complicated, approximations in conventional differentiable manifolds (Pohl, 1962).

Indeed, Rate Distortion Manifolds can be quite formally described using standard techniques from topological manifold theory (Glazebrook and Wallace, 2007). The essential point is that a rate distortion manifold is a topological structure which constrains the multifactorial stream of high level cognition as well as the pattern of communication between giant components, much the way a riverbank constrains the flow of the river it contains. This is a fundamental insight, which we will pursue further.

The Rate Distortion Manifold can, however, also be described in purely information theoretic terms using a 'tuning theorem' variant of the Shannon Coding Theorem, which we briefly review.

Messages from an information source, seen as symbols x_j from some alphabet, each having probabilities P_j associated with a random variable X, are 'encoded' into the language of a 'transmission channel', a random variable Y with symbols y_k, having probabilities P_k, possibly with error. Someone receiving the symbol y_k then retranslates it (without error) into some x_k, which may or may not be the same as the x_j that was sent.

More formally, the message sent along the channel is characterized by a random variable X having the distribution

$$P(X = x_j) = P_j, j = 1, ..., M.$$

The channel through which the message is sent is characterized by a second random variable Y having the distribution

$$P(Y = y_k) = P_k, k = 1, ..., L.$$

Let the joint probability distribution of X and Y be defined as

$$P(X = x_j, Y = y_k) = P(x_j, y_k) = P_{j,k}$$

and the conditional probability of Y given X as

$$P(Y = y_k | X = x_j) = P(y_k | x_j).$$

Then the Shannon uncertainty of X and Y independently and the joint uncertainty of X and Y together are defined respectively as

$$H(X) = -\sum_{j=1}^{M} P_j \log(P_j)$$

$$H(Y) = -\sum_{k=1}^{L} P_k \log(P_k)$$

$$H(X,Y) = -\sum_{j=1}^{M}\sum_{k=1}^{L} P_{j,k} \log(P_{j,k}).$$

(2.21)

The *conditional uncertainty* of Y given X is defined as

$$H(Y|X) = -\sum_{j=1}^{M}\sum_{k=1}^{L} P_{j,k} \log[P(y_k|x_j)].$$

(2.22)

For any two stochastic variates X and Y, $H(Y) \geq H(Y|X)$, as knowledge of X generally gives some knowledge of Y. Equality occurs only in the case of stochastic independence.

Since $P(x_j, y_k) = P(x_j)P(y_k|x_j)$, we have

$$H(X|Y) = H(X,Y) - H(Y).$$

The information transmitted by translating the variable X into the channel transmission variable Y – possibly with error – and then retranslating without error the transmitted Y back into X is defined as

$$I(X|Y) \equiv H(X) - H(X|Y) = H(X) + H(Y) - H(X,Y)$$

(2.23)

See, for example, Ash (1990), Khinchin (1957) or Cover and Thomas (1991) for details. The essential point is that if there is no uncertainty in X given the channel Y, then there is no loss of information through transmission. In general this will not be true, and herein lies the essence of the theory.

Given a fixed vocabulary for the transmitted variable X, and a fixed vocabulary and probability distribution for the channel Y, we may vary the probability distribution of X in such a way as to maximize the information sent. The capacity of the channel is defined as

$$C \equiv \max_{P(X)} I(X|Y)$$

(2.24)

subject to the subsidiary condition that $\sum P(X) = 1$.

The critical trick of the Shannon Coding Theorem for sending a message with arbitrarily small error along the channel Y at any rate $R < C$ is to encode it in longer and longer 'typical' sequences of the variable X; that is, those sequences whose distribution of symbols approximates the probability distribution $P(X)$ above which maximizes C.

If $S(n)$ is the number of such 'typical' sequences of length n, then

$$\log[S(n)] \approx nH(X),$$

where $H(X)$ is the uncertainty of the stochastic variable defined above. Some consideration shows that $S(n)$ is much less than the total number of possible messages of length n. Thus, as $n \to \infty$, only a vanishingly small fraction of all possible messages is meaningful in this sense. This observation, after some considerable development, is what allows the Coding Theorem to work so well. In sum, the prescription is to encode messages in typical sequences, which are sent at very nearly the capacity of the channel. As the encoded messages become longer and longer, their maximum possible rate of transmission without error approaches channel capacity as a limit. Again, Ash (1990), Khinchin (1957) and Cover and Thomas (1991) provide details.

This approach can be, in a sense, inverted to give a tuning theorem which parsimoniously describes the essence of the Rate Distortion Manifold.

Telephone lines, optical wave, guides and the tenuous plasma through which a planetary probe transmits data to earth may all be viewed in traditional information-theoretic terms as a *noisy channel* around which we must structure a message so as to attain an optimal error-free transmission rate.

Telephone lines, wave guides, and interplanetary plasmas are, relatively speaking, fixed on the timescale of most messages, as are most other signaling networks. Indeed, the capacity of a channel, is defined by varying the probability distribution of the 'message' process X so as to maximize $I(X|Y)$.

Suppose there is some message X so critical that its probability distribution must remain fixed. The trick is to fix the distribution $P(x)$ but *modify the channel* – i.e., tune it – so as to maximize $I(X|Y)$. The *dual* channel capacity C^* can be defined as

$$C^* \equiv \max_{P(Y),P(Y|X)} I(X|Y).$$

(2.25)

But

$$C^* = \max_{P(Y),P(Y|X)} I(Y|X)$$

since

$$I(X|Y) = H(X) + H(Y) - H(X,Y) = I(Y|X).$$

Thus, in a purely formal mathematical sense, *the message transmits the channel*, and there will indeed be, according to the Coding Theorem, a channel distribution $P(Y)$ which maximizes C^*.

One may do better than this, however, by modifying the channel matrix $P(Y|X)$. Since

$$P(y_j) = \sum_{i=1}^{M} P(x_i)P(y_j|x_i),$$

$P(Y)$ is entirely defined by the channel matrix $P(Y|X)$ for fixed $P(X)$ and

$$C^* = \max_{P(Y),P(Y|X)} I(Y|X) = \max_{P(Y|X)} I(Y|X).$$

Calculating C^* requires maximizing the complicated expression

$$I(X|Y) = H(X) + H(Y) - H(X,Y),$$

which contains products of terms and their logs, subject to constraints that the sums of probabilities are 1 and each probability is itself between 0 and 1. Maximization is done by varying the channel matrix terms $P(y_j|x_i)$ within the constraints. This is a difficult problem in nonlinear optimization. However, for the special case $M = L$, C^* may be found by inspection: If $M = L$, then choose

$$P(y_j|x_i) = \delta_{j,i},$$

where $\delta_{i,j}$ is 1 if $i = j$ and 0 otherwise. For this special case

$$C^* \equiv H(X),$$

with $P(y_k) = P(x_k)$ for all k. *Information is thus transmitted without error when the channel becomes 'typical' with respect to the fixed message distribution $P(X)$.*

If $M < L$, matters reduce to this case, but for $L < M$ information must be lost, leading to Rate Distortion limitations.

Thus modifying the channel may be a far more efficient means of ensuring transmission of an important message than encoding that message in a 'natural' language which maximizes the rate of transmission of information on a fixed channel.

We have examined the two limits in which either the distributions of $P(Y)$ or of $P(X)$ are kept fixed. The first provides the usual Shannon Coding Theorem, and the second, hopefully, a tuning theorem variant, a tunable retina-like Rate Distortion Manifold. It seems likely, however, than for many important systems $P(X)$ and $P(Y)$ will interpenetrate, to use Richard Levins' terminology. That is, $P(X)$ and $P(Y)$ will affect each other in characteristic ways, so that some form of mutual tuning may be the most effective strategy.

2.11 No free lunch

The previous results can be used to give yet another perspective on the famous 'no free lunch' theorem of Wolpert and Macready (1995, 1997). As English (1996) states the matter,

"...Wolpert and Macready... have established that there exists no generally superior function optimizer. There is no 'free lunch' in the sense that an optimizer 'pays' for superior performance on some functions with inferior performance on others... if the distribution of functions is uniform, then gains and losses balance precisely, and all optimizers have identical average performance... The formal demonstration depends primarily upon a theorem that describes how information is conserved in optimization. This Conservation Lemma states that when an optimizer evaluates points, the posterior joint distribution of values for those points is exactly the prior joint distribution. Put simply, observing the values of a randomly selected function does not change the distribution...

[A]n optimizer has to 'pay' for its superiority on one subset of functions with inferiority on the complementary subset...

Anyone slightly familiar with the [evolutionary computing] literature recognizes the paper template 'Algorithm X was treated with modification Y to obtain the best known results for problems P_1 and P_2.' Anyone who has tried to find subsequent reports on 'promising' algorithms knows that they are extremely rare. Why should this be?

A claim that an algorithm is the very best for two functions is a claim that it is the very worst, on average, for all but two functions.... It is due to the diversity of the benchmark set [of test problems] that the 'promise' is rarely realized. Boosting performance for one subset of the problems usually detracts from performance for the complement...

Hammers contain information about the distribution of nail-driving problems. Screwdrivers contain information about the distribution of screw-driving problems. Swiss army knives contain information about a broad distribution of survival problems. Swiss army knives do many jobs, but none particularly well. When the many jobs must be done under primitive conditions, Swiss army knives are ideal.

The tool literally carries information about the task... optimizers are literally tools-an algorithm implemented by a computing device is a physical entity..."

Another way of stating this conundrum is to say that a computed solution is simply the product of the information processing of a problem, and, by a very famous argument, information can never be gained simply by processing. Thus a problem X is transmitted as a message by an information processing channel, Y, a computing device, and recoded as an answer. By the 'dual' argument of the previous section there will be a channel coding of Y which, when properly tuned, is most efficiently transmitted by the problem. In general, then, the most efficient coding of the transmission channel, that is, the best algorithm turning a problem into a solution, will necessarily be highly problem-specific. Thus there can be no best algorithm for all sets of problems, although there may well be an optimal algorithm for any given set.

Rate distortion, however, occurs when the problem is collapsed into a smaller, simplified version and then solved. Then there must be a tradeoff between allowed average distortion and the rate of solution: the retina effect. In a very fundamental sense, then, Rate Distortion Manifolds present a generalization of the converse of the Wolpert/Macready no free lunch arguments.

An important point is that this development says nothing about the efficiency of channel tuning. Another iteration of theory would be required to determine, say, the full width at half maximum (FWHM) of a particular system. The peak in dual channel capacity may be very broad in any given case, so that although one algorithm (or equivalence class of them) may be 'best', very many others may be nearly as good. One suspects that for a Rate Distortion Manifold/retina system tunability might sometimes be more critical.

2.12 Mesoscale resonance: Many Baldwin effects

These arguments lead, in turn, to a generalized form of stochastic resonance applicable to a spectrum of phenomena similar to the Simpson-Baldwin effect. Ancel (1999) has presented what is perhaps the clearest mathematical model of the basic idea.

She argues that organisms that make non-hereditary physical or behavioral modifications to survive environmental stresses will have better representation in future generations than less versatile organisms. Through natural selection, then, the capacity for such adaptation along with the beneficial acquired traits will become universal. Calculation shows that the distribution of phenotypes in a population depends largely on the extent of environmental stochasticity. When the environment undergoes intermediate rates of fluctuation, the Baldwin effect arises through the interaction of natural selection and mutation on norms of reaction. In a highly volatile environment by contrast, organisms benefit from plasticity, and consequently do not experience a Simpson-Baldwin channeling of phenotype possibility.

The essential point, from our perspective, is the importance of intermediate rates of environmental fluctuation, which mirrors the generalized stochastic resonance arguments we ultimately invoke.

Classic stochastic resonance (SR) emerges from the arguments of the previous sections quite directly. The only coding possible under the conditions of SR is to add random noise to the amplitude of a structured signal which, by itself, is below threshold for triggering some powerful, highly nonlinear device. The only 'tuning' possible to random noise is to vary its amplitude. By the arguments above, there will be some optimum noise amplitude which will maximize the dual channel capacity, and hence the transmission rate of the signal via the powerful, threshold-driven, oscillator.

Similarly, in Ancel's model of the Baldwin effect, the only 'tuning' possible the system, as she has presented it, is the extent of environmental stochasticity.

By these arguments, then, there will be an 'optimal' level of stochasticity driving the effect.

Clearly more complicated natural processes can be subject to analogous tuning, with particular sensitivity to 'intermediate' level effects, effectively a mesoscale resonance in which ecosystem changes entrain phenomena at other scales. This is a critical point, and on it will subsequently hang much of our tale.

Again, something much like this result has been reached by others via different means. Gavrilets (2003), for example, describes numerical individual-oriented population genetics models as producing similar outcomes:

> "Theoretical studies predict extreme sensitivity of the probability of speciation and the waiting time to speciation on model parameters which in turn strongly depend on the environmental conditions. This suggests that in general speciation is triggered by changes in the environment."

Our contribution is to provide an explicit analytic framework for this effect.

2.13 Directed homotopy

To reiterate, the groupoid treatment of modular cognitive networks above defined equivalence classes of states according to whether they could be linked by grammatical/syntactical high-probability 'meaningful' paths. The dynamical groupoid is based on identification of equivalence classes of languages.

Next we ask the precisely complementary question regarding paths on dynamical manifolds: For any two particular given states, is there some sense in which we can define equivalence classes across the set of meaningful paths linking them? The assumption is that the system has been 'tuned' using the Rate Distortion Manifold approach above, so that the problem to be solved is more tractable, in a sense.

This is of particular interest to the second order hierarchical model of the next chapter which, in effect, describes a universality class tuning of the renormalization parameters characterizing the dancing, flowing, tunably punctuated accession to high order cognitive function.

A closely similar question is central to recent algebraic geometry approaches to concurrent, that is, highly parallel, computing (Pratt, 1991; Goubault and Raussen, 2002; Goubault, 2003), which we adapt.

For the moment we restrict the analysis to a giant component system characterized by two Morse-theoretic renormalization parameters, say ω_1 and ω_2, and consider the set of meaningful paths connecting two particular points, say a and b, in the two dimensional ω-space plane of figure 2.1. The arguments surrounding equations (2.7), (2.8) and (2.13) suggests that there may be regions of fatal attraction and strong repulsion, Black holes and White holes, which can either trap or deflect the path of multitasking cognition.

Figures 2.1 and 2.2 show two possible configurations for a Black and a White hole, diagonal and cross-diagonal. If one requires path monotonicity – always increasing or remaining the same – then, following Goubault (2003, figs. 6,7), there are, intuitively, two direct ways, without switchbacks, that one can get from a to b in the diagonal geometry of figure 2.1, without crossing a Black or White hole. But there are three in the cross-diagonal structure of figure 2.2.

Elements of each 'way' can be transformed into each other by continuous deformation without crossing either the Black or White hole. Figure 2.1 has two additional possible monotonic ways, involving over/under switchbacks, which are not drawn. Relaxing the monotonicity requirement generates a plethora of other possibilities, for example, loopings and backwards switchbacks, whose consideration is left as an exercise. It is not clear under what circumstances such complex paths can be meaningful, a matter for further study.

These ways are the equivalence classes defining the topological structure of the two different ω-spaces, analogs to the fundamental homotopy groups in spaces which admit of loops (Lee, 2000). The closed loops needed for classical homotopy theory are impossible for this kind of system because of the 'flow of time' defining the output of an information source – one goes from a to b although, for nonmonotonic paths, intermediate looping would seem possible. The theory is thus one of directed homotopy, dihomotopy, and the central question revolves around the continuous deformation of paths in ω-space into one another, without crossing Black or White holes. Goubault and Rausssen (2002) provide another introduction to the formalism.

These ideas can, of course, be applied to lower level cognitive modules as well as to the second order hierarchical cognitive model where they are, perhaps, of more central interest.

It seems likely that external signals or developmental history can define quite different dihomotopies of attentional focus in cognitive gene expression or other similar phenomena. That is, the topology of blind spots will be developmentally modulated. It is this ontogenetic topology of multitasking attention which, acting in concert with the inherent limitations of the rate distortion manifold, generates the pattern of overfocus analogous to inattentional blindness.

Such considerations, and indeed the Black Hole development of equation (2.13), suggest that a multitasking system which becomes trapped in a particular pattern cannot, in general, expect to emerge from it in the absence of external forcing mechanisms or the stochastic resonance/mutational action of 'noise'. Emerging from such a trap involves large-scale topological changes, and this is the functional equivalent of a first order phase transition in a physical systems which requires energy to overcome latent heat.

A second perspective is that the solution to what is essentially a kind of highly parallel computing problem might well be cast in terms of 'finding the right Black Hole'. That is, the system begins at some starting point, and,

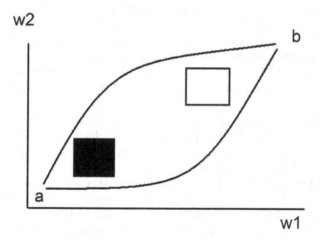

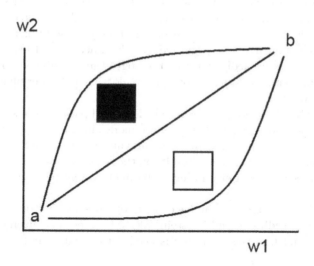

Fig. 2.1. Diagonal Black and White holes in the two dimensional ω-plane. Only two direct paths can link points a and b which are continuously deformable into one another without crossing either hole. There are two additional monotonic switchback paths which are not drawn.

Fig. 2.2. Cross-diagonal Black and White holes. Three direct equivalence classes of continuously deformable paths can link a and b. Thus the two spaces are topologically distinct. Here monotonic switchbacks are not possible, although relaxation of that condition can lead to 'backwards' switchbacks and intermediate loopings.

if the environment has effectively constrained the shape of the underlying dynamic manifold, that is, to act as a tuning 'goal context' in Baars' terminology, a solution involves identifying the various topological modes possible to that complicated geometric object. Among those modes will be the absorbing singularities which constitute an analog to problem solution, although convergence is not to a fixed state, but rather to some set of highly dynamic information sources.

The fundamental insight is that the equivalence class structure of environmental context can define a groupoid symmetry which can then be imposed on the 'natural' groupoids underlying massively parallel gene expression.

This sort of behavior is, as we have noted, central to ecosystem resilience theory (Gunderson, 2000; Holling, 1973). The essential idea is that equivalence classes of dynamic manifolds, and the directed homotopy classes within those manifolds, each and together create domains of quasi-stability requiring action of some external signal for change. We will explore this in more detail in some of the following chapters.

Apparently the set of dynamic manifolds, and its subsets of directed homotopy equivalence classes, formally classifies quasi-equilibrium states, and thus characterizes the different possible resilience modes. Some of these may be highly pathological. Others, however, will represent 'solutions' to the gene expression (or other dynamic cognitive) problem, according to this scheme.

Shifts between markedly different topological modes appear to be necessary effects of phase transitions, involving analogs to first order phase changes in physical systems.

It seems clear that both 'problem solutions' and pathological states can be represented as topological resilience/phase modes in this model, suggesting a real equivalence between difficulties in carrying out gene expression and its stabilization. This mirrors recent results on the relation between programming difficulty and system stability in highly parallel computing devices (Wallace, 2008a).

Some 99% of the human genome consists of sequence that does not encode proteins (e.g., Sasidharan and Gerstein, 2008). Our results converge with conjectures that much of this structure is needed to stabilize and regulate gene expression.

2.14 Pathologies of gene expression

Something of this ambiguity can be explicitly modeled.

Suppose we can operationalize and quantify degrees of both inattentional blindness (IAB) and of average distortion (D) in communication between gene expression networks and signals from an embedding context. D and IAB are thus taken as contexts in the Baars' sense, for cognitive gene expression. The essential assumption is that the dual information source of a collective which has low levels of both IAB and D will tend to be richer than that of one having greater levels. This is shown in figure 2.3a, where H is the source uncertainty, $X = IAB$, and $Y = D$. Regions of low X, Y, near the origin, have greater source uncertainty than those nearby, so $H(X, Y)$ shows a (relatively gentle) peak at the origin, taken here as simply the product of two error functions.

The generalized Onsager argument of equations 2.7-2.14 is shown in figure 2.3b, where $S = H(X, Y) - X dH/dX - Y dH/dY$ is graphed on the Z axis against the $X - Y$ plane, assuming a gentle peak in H at the origin. Peaks in S, according to theory, constitute repulsive system barriers, which must be overcome by external forces. In figure 2.3b there are three quasi-stable topological resilience modes, marked as A, B, and C. The A region is locked in to low levels of both inattentional blindness and average distortion, as it sits in a pocket. Forcing the system in either direction, that is, increasing either IAB or D, will, initially, be met by homeostatic attempts to return to the resilience state A, according to this model.

If, in particular, rate distortion problems become severe in spite of homeostatic mechanisms, the system will then jump to the quasi-stable state B, a second pocket. According to the model, however, once that transition takes place, there will be a tendency for the system to remain in a condition of high average distortion. That is, the system will, according to the model, become locked in to a structure with high distortion in communication between gene expression and embedding context, but one having lower overall collective capacity, a lower value of H in figure 2.3a.

The third pocket, marked C, is a broad plain in which both IAB and D remain high, a highly overfocused, poorly crosslinked structure which will require significant intervention to alter once it reaches such a quasi-stable resilience mode. Collective capacity, measured by H in figure 2.3a, is the lowest of all for this condition of pathological resilience, and attempts to correct the problem – to return to condition A will be met with very high barriers in S, according to figure 2.3b. That is, mode C is very highly resilient, although pathologically so, much like the eutrophication of a pure lake by sewage outflow.

We can argue that the three quasi-equilibrium configurations of figure 2.3b represent different dynamical states of the system, and that the possibility of transition between them represents the breaking of the associated symmetry groupoid by external forcing mechanisms. That is, three manifolds representing three different kinds of system dynamics have been patched together to create a more complicated topological structure. For cognitive phenomena, this phenomenon is likely to be the rule rather than the exception. 'Pure' groupoids are likely to be arbitrary abstractions, and the fundamental questions will involve the systems of linkages which break the underlying symmetry.

In all this, of course, 'convergence' is not to some fixed state, limit cycle, or pseudorandom strange attractor, but rather to some appropriate set of information sources, which is a very different thing indeed.

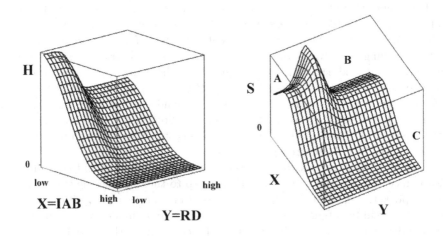

S=H-XdH/dX-YdS/dY

Fig. 2.3. a. Source uncertainty, H, of the dual information source of multiple workspace gene expression, as parametized by degrees of inattentional blindness, $X = IAB$ and average distortion $Y = D$, taken here as environmental indices of embedding contexts. Note the relatively gentle peak at low values of X, Y. H is generated as the product of two error functions. b. Generalized Onsager treatment of figure 2.3a. $S = H(X, Y) - XdH/dX - YdH/dY$. The regions marked A, B, and C represent realms of resilient quasi-stability, divided by barriers defined by the relative peaks in S. Transition among them requires a forcing mechanism. From another perspective, limiting resources or imposing stress from the outside – driving down H in figure 2.3a, would force the system into the lower plain of C, in which the system would then become trapped in states having high levels of average distortion and inattentional blindness.

What this model suggests is that sufficiently strong external perturbation can force gene expression networks within an organism from a normal development path into one involving widespread, comorbid, developmental disorder. This is a widely studied pattern for humans (e.g., Wallace and Fullilove, 2008). Indeed, the results of this section might well serve as the foundation of a fairly comprehensive model of chronic developmental disease. One approach might be as follows:

A developmental process can be viewed as involving a sequence of surfaces like figure 2.3, having, for example, 'critical periods' of high energy demand when the barriers between the normal state A and the pathological states B and C are relatively low. During such a time the system would become highly sensitive to perturbation, and to the onset of a subsequent pathological developmental trajectory. Critical periods might occur during times of rapid growth for which an energy limitation imposes the need to focus via something like a Rate Distortion Manifold. Cognitive process requires energy through the homology between equations (1.5) and (2.4), and more focus at one end necessarily implies less at some other. In a distributed zero sum developmental game, as it were, some cognitive processes must receive more attentional energy than others.

Other mechanisms might well operate in a similar manner. For example the impact of long-term psychosocial or other environmental stressors could induce critical periods in which exposure to acute noxious conditions could then trigger pathological developmental trajectories of gene expression.

2.15 Traveling waves on cognitive modular networks

Sections 2.8 and 2.13 above explored two kinds of manifolds on modular information networks. Here we introduce yet another.

It is important to remember that we cannot easily observe modular networks dual to cognitive process directly, but must often infer what we can from examining the messages they send. One important characteristic is simply the rate of transmission. Can we determine something of network structure from rate information?

Suppose that the probability of interaction between individual nodes in a network depends jointly on (at least) two distinct measures which we characterize as multidimensional vector quantities $\mathbf{X}$ and $\mathbf{Z}$ respectively. These might be determined by a variety of empirical means.

For individual nodes j and k we assume their probability of interaction $P_{j,k}$ is given by

$$P_{j,k} = P_{j,k}(\mathbf{X}_j, \mathbf{X}_k, \mathbf{Z}_j, \mathbf{Z}_k),$$

where $0 \leq P_{j,k} \leq 1$.

Assume it possible to reduce $P_{j,k}$ to a function of the differences $\mathbf{X} = \mathbf{X}_j - \mathbf{X}_k$ and $\mathbf{Z} = \mathbf{Z}_j - \mathbf{Z}_k$, or, perhaps, by using a multivariate method such as principal components analysis, even to functions of their 'length' $x = |\mathbf{X}|$ and $z = |\mathbf{Z}|$, so that

$$P_{j,k} = P_{j,k}(x, z).$$

One way to proceed is to impose a generalized distance $r^2 \equiv x^2 + z^2$, and explore the effects of various probability distributions which are functions of r. Here, rather, we finesse the argument and transform out of the space defined by x and z into the probability space itself, defining a metric according to

$$L_{j,k} \equiv \log(1/P_{j,k}),$$

(2.26)

where log is the logarithm to some base number. Note that it is the probability distribution based on the generalized distance r which induces the transformation between 'real' space and probability space.

If it can be assumed for all nodes j, k, l within a sufficiently small network patch – the essence of a manifold – that

$$P_{j,l} \geq P_{j,k}P_{k,l},$$

so that

$$\frac{1}{P_{j,l}} \leq \frac{1}{P_{j,k}}\frac{1}{P_{k,l}}$$

and the strong 'triangle' inequality

$$\log(1/P_{j,l}) \leq \log(1/P_{j,k}) + \log(1/P_{k,l})$$

holds, then L is a pseudometric, and various standard attacks are possible.

That somewhat draconian condition can, however, be considerably weakened as follows:

Let ΔL_j be the average 'distance' in probability space from the node j to all other nodes, that is,

$$\Delta L_j \equiv \sum_k P_{j,k} \log(1/P_{j,k}).$$

(2.27)

Suppose some fairly elaborate 'message,' not otherwise described, is sent along the network, and a *traveling wave* condition is imposed, so that, for some time period Δt, the relation

$$\frac{\Delta L_j}{\Delta t} \approx C,$$

(2.28)

holds. C is, then, *the mean fixed rate at which the message is sent from the network to an embedded individual.*

We will subsequently show – not unexpectedly and probably not originally – that the traveling wave assumption, on a fractal manifold, is equivalent to the Aharony-Stauffer conjecture (Aharony and Stauffer, 1984). The conjecture directly relates the fractal dimension of the interior of an affected network to that of its growth surface. This gives a simple and explicit expression for C, usually an arduous calculation.

By expanding equation (2.28) we obtain, taking $\Delta t \equiv 1$,

$$-\sum_k P_{j,k} \log(P_{j,k}) \approx C,$$

(2.29)

where C is a transmission rate constant characteristic of the particular sociogeographic network. We further assume that $\sum_k P_{j,k} = 1$, i.e., the network is 'tight' in the sense that each node interacts with it as a whole with unit probability. Hence $P_{j,k}$ is a legitimate probability distribution.

For any probability distribution, $0 \le P_j \le 1$ such that $\sum_j P_j = 1$ the quantity $H = -\sum_j P_j \log(P_j)$ is, of course, the distribution's Shannon uncertainty.

The transfer of uncertainty represents the transmission of information. The Shannon Coding Theorem states that for any rate $R < C$, where C represents the capacity of the information channel, it is possible to find a 'coding scheme' such that a sufficiently long message can be sent with arbitrarily small error.

We are interested in network manifolds upon which it is possible to define an information metric as above. Heuristically, we take $P_{i,j}$ to be the joint probability that something of the form x_j is encoded and sent from the point j in the network and received as something of the form y_i at the point i. $P_{i,j}$ thus jointly characterizes the interaction of a network and an individual within it.

Again define the 'average distance' between i and the network as in equation (2.27), taking $\Delta L_i = -\sum_j P_{i,j} \log(P_{i,j})$, and assume that to be less than the 'radius' of the neighborhood mapped homogeneously onto some more regular network. We may need to assume $P_{i,j} \neq 0$ only for i and j within a homogeneously mapped patch.

Over that patch

$$\sum_i \Delta L_i = H(X,Y),$$

where $H(X,Y)$ is precisely the *joint uncertainty* of an interaction between a network and each individual within it. Let X represent 'those who know and tell' within the homogeneous patch, and Y 'those who do not know and listen', having probability distributions P_i and P_j, and uncertainties based on these distributions $H(X)$ and $H(Y)$. The transmission of a message within the network can be described as the transmission between X and Y, i.e, between those who know and tell and those who do not know and listen,

$$I(X|Y) = H(X) + H(Y) - H(X,Y)$$

which can take place at a maximum rate defined by the capacity of the channel connecting the knowers and the ignorant.

The point of all this is that metrics defined by $\log(1/P)$, where P is a probability, indeed have essential information theory meanings.

Assume that the network of interest is 'scale free' in an information metric L, so that the number of nodes within a distance L of a given individual, $N(L)$, has the form

$$N(L) \approx \mu\rho L^{D_F}$$

(2.30)

where ρ is the analog of a density of individuals per unit area, D_F is a nonnegative real number, and μ a scaling factor.

Assume a region of radius L from a given individual node and containing $N(L)$ other nodes is sent a message from that first node, and that the region receiving the message grows in time by communicating with other nodes near its boundary according to another power law

$$dN/dt \approx g(\rho)L^{D_G}$$

(2.31)

where t is the time, $g(\rho)$ is a growth function of network density ρ, and D_G is a nonnegative real exponent which characterizes the growth surface of the transmitting region. That growth surface is taken as having a small thickness compared to the characteristic radius L of the transmitting region. D_G is not in general the same as D_F, and neither may be an integer, i.e., this can be a 'fractal', scale-free phenomenon. Differentiating the first of these equations and substituting the result into the left hand side of the second gives

$$C \equiv dL/dt \approx \frac{g(\rho)L^{D_G - D_F + 1}}{\mu\rho D_F}.$$

(2.32)

Imposing a traveling wave solution to this expression – the assumption that dL/dt is independent of L – is equivalent to having $D_G = D_F - 1$. This is the Aharony-Stauffer conjecture for fractal surfaces. Then

$$C = dL/dt = \frac{g(\rho)}{\mu\rho D_F}$$

(2.33)

is a system constant depending on both ρ and D_F. It represents, for our particular choice of L, a kind of patch-wise local channel capacity for the network.

Typically $g(\rho)$ can be approximated by some polynomial in ρ having a dominant term $k + 1$. Thus equation (2.33) becomes, in first order,

$$C \propto \rho^k / D_F.$$

(2.34)

While we may not often directly observe modular cognitive networks, we can observe both the messages they send and the rate of transmission. Equation (2.34) relates certain of their local topological properties to the local rate of transmission. Changes in local transmission rate, in this model, are sufficient to imply changes in local topology.

3

Formal theory II

3.1 The mean field model

This chapter outlines a somewhat different approach to the problem of tunable phase transitions in cognitive and other systems characterized by information sources. Here we use a 'mean field' approximation in contrast to the preceding chapter which can be considered a 'mean number' treatment. Hybrid models also seem possible.

Wallace and Wallace (1998, 1999) addressed how a language, in the broadest sense, 'spoken' on a network structure, responds as properties of the network change. The language might be speech, pattern recognition, or cognition. The network might be social, chemical, or neural. The properties of interest were the magnitude of 'strong' or 'weak' ties which, respectively, either disjointly partitioned the network or linked it across such partitioning. These would be analogous to local and mean-field couplings in physical systems. Here the system of interest is a set of information sources dual to cognitive processes which becomes linked through an average coupling by crosstalk.

Fix the magnitude of strong ties – again, those which disjointly partition the underlying network into cognitive or other submodules – but vary the index of nondisjunctive weak ties, P, between components, taking $K = 1/P$.

Assume the piecewise, adiabatically stationary ergodic information source depends on three parameters, two explicit and one implicit. The explicit are K as above and, as a calculational device, an 'external field strength' analog J, which gives a 'direction' to the system. We will, in the limit, set $J = 0$. Note that there are many other ways of doing this, since classical renormalization techniques are more philosophy than prescription.

The implicit parameter, r, is an inherent generalized 'length' characteristic of the phenomenon, on which J and K are defined. That is, J and K are written as functions of averages of the parameter r, which may be quite complex, having nothing at all to do with conventional ideas of space. For example, r may be defined by the degree of niche partitioning in ecosystems or separation in social structures, and similarly for biological networks of various kinds.

R. Wallace et al., *Farming Human Pathogens*, DOI 10.1007/978-0-387-92213-3_3,
© Springer Science+Business Media, LLC 2009

For a given generalized language of interest having a well defined (adiabatically, piecewise stationary) ergodic source uncertainty, $H = H[K, J, \mathbf{X}]$.

To summarize a long train of standard argument (Binney et al., 1986; Wilson, 1971), imposition of invariance of H under some renormalization transform in the implicit parameter r leads to expectation of both a critical point in K, written K_C, reflecting a phase transition to or from collective behavior across the entire array, and of power laws for system behavior near K_C. Addition of other parameters to the system results in a 'critical line' or surface.

Let $\kappa \equiv (K_C - K)/K_C$ and take χ as the 'correlation length' defining the average domain in r-space for which the information source is primarily dominated by 'strong' ties. The first step is to average across r-space in terms of 'clumps' of length $R =< r >$. Then $H[J, K, \mathbf{X}] \rightarrow H[J_R, K_R, \mathbf{X}]$.

Taking Wilson's (1971) analysis as a starting point – not the only way to proceed – the 'renormalization relations' used here are:

$$H[K_R, J_R, \mathbf{X}] = f(R)H[K, J, \mathbf{X}]$$

$$\chi(K_R, J_R) = \frac{\chi(K, J)}{R},$$

(3.1)

with $f(1) = 1$ and $J_1 = J, K_1 = K$. The first equation significantly extends Wilson's treatment. It states that 'processing capacity,' as indexed by the source uncertainty of the system, representing the 'richness' of the generalized language, grows monotonically as $f(R)$, which must itself be a dimensionless function in R, since both $H[K_R, J_R]$ and $H[K, J]$ are themselves dimensionless. Most simply, this requires replacing R by R/R_0, where R_0 is the 'characteristic length' for the system over which renormalization procedures are reasonable, then setting $R_0 \equiv 1$. Length is measured in units of R_0.

Wilson's original analysis focused on free energy density. Under 'clumping,' densities must remain the same, so that if $F[K_R, J_R]$ is the free energy of the clumped system, and $F[K, J]$ is the free energy density before clumping, then Wilson's equation (4) is $F[K, J] = R^{-3}F[K_R, J_R]$, so that

$$F[K_R, J_R] = R^3 F[K, J].$$

Remarkably, the renormalization equations are solvable for a broad class of functions $f(R)$, or more precisely, $f(R/R_0), R_0 \equiv 1$.

The second equation just states that the correlation length simply scales as R.

Again, the central feature of renormalization in this context is the assumption that, at criticality, the system looks the same at all scales, that is, it is *invariant under renormalization* at the critical point. All else flows from this.

There is no unique renormalization procedure for information sources. Other, very subtle, symmetry relations – not necessarily based on the elementary physical analog we use here – may well be possible. For example, McCauley (1993, p.168) describes the highly counterintuitive renormalizations needed to understand phase transition in simple 'chaotic' systems. This is important, since biological or social systems may well alter their renormalization properties – equivalent to tuning their phase transition dynamics – in response to external signals. We will make much use of a simple version of this possibility, termed 'universality class tuning,' below.

To begin, following Wilson, take $f(R) = R^d$, d some real number $d > 0$, and restrict K to near the 'critical value' K_C. If $J \to 0$, a simple series expansion and some clever algebra (Wilson, 1971; Binney et al., 1986) gives

$$H = H_0 \kappa^\alpha$$

$$\chi = \frac{\chi_0}{\kappa^s},$$

(3.2)

where α, s are positive constants. Other, more biologically relevant, examples appear below, and will be used to construct a higher level dynamical groupoid in the sense of section 2.6 above.

Further from the critical point, matters are more complicated, appearing to involve Generalized Onsager Relations, 'dynamical groupoids', and a kind of thermodynamics associated with a Legendre transform of H: $S \equiv H - K dH/dK$.

An essential insight is that *regardless of the particular renormalization properties, sudden critical point transition is possible in the opposite direction for this model.* That is, one goes from a number of independent isolated and fragmented systems operating individually and more or less at random into a single large, interlocked, coherent structure, once the parameter K, the inverse strength of weak ties, falls below threshold, or, conversely, once the strength of weak ties parameter $P = 1/K$ becomes large enough.

Thus, increasing nondisjunctive weak ties between them can bind several different cognitive 'language' functions into a single, embedding hierarchical metalanguage containing each as a linked subdialect, and do so in an inherently punctuated manner. This could be a dynamic process, creating a shifting, ever-changing pattern of linked cognitive submodules, according to the challenges or opportunities faced by the organism.

This heuristic insight can be made more exact using a rate distortion argument (or, more generally, using the Joint Asymptotic Equipartition Theorem) as follows:

Suppose that two ergodic information sources **Y** and **B** begin to interact, to 'talk' to each other, to influence each other in some way so that it is possible, for example, to look at the output of **B** – strings b – and infer something about the behavior of **Y** from it – strings y. We suppose it possible to define a retranslation from the B-language into the Y-language through a deterministic code book, and call $\hat{\mathbf{Y}}$ the translated information source, as mirrored by **B**.

Define some distortion measure comparing paths y to paths $\hat{y}$, $d(y, \hat{y})$. Invoke the Rate Distortion Theorem's mutual information $I(Y, \hat{Y})$, which is the splitting criterion between high and low probability pairs of paths. Impose, now, a parametization by an inverse coupling strength K, and a renormalization representing the global structure of the system coupling. This may be much different from the renormalization behavior of the individual components. If $K < K_C$, where K_C is a critical point (or surface), the two information sources will be closely coupled enough to be characterized as condensed.

In the absence of a distortion measure, the Joint Asymptotic Equipartition Theorem gives a similar result.

Detailed coupling mechanisms will be sharply constrained through regularities of grammar and syntax imposed by limit theorems associated with phase transition.

3.2 Biological renormalization

Next, the mathematical detail concealed by the invocation of the asymptotic limit theorems emerges with a vengeance. Equation (3.1) states that the information source and the correlation length, the degree of coherence on the underlying network, scale under renormalization clustering in chunks of size R as

$$H[K_R, J_R]/f(R) = H[J, K]$$

$$\chi[K_R, J_R]R = \chi(K, J),$$

with $f(1) = 1, K_1 = K, J_1 = J$, where we have slightly rearranged terms.

Differentiating these two equations with respect to R, so that the right hand sides are zero, and solving for dK_R/dR and dJ_R/dR gives, after some consolidation, expressions of the form

$$dK_R/dR = u_1 d\log(f)/dR + u_2/R$$

$$dJ_R/dR = v_1 J_R d\log(f)/dR + \frac{v_2}{R} J_R.$$

(3.3)

The $u_i, v_i, i = 1, 2$ are functions of K_R, J_R, but not explicitly of R itself.
We expand these equations about the critical value $K_R = K_C$ and about $J_R = 0$, obtaining

$$dK_R/dR = (K_R - K_C)yd\log(f)/dR + (K_R - K_C)z/R$$

$$dJ_R/dR = wJ_R d\log(f)/dR + xJ_R/R.$$

(3.4)

The terms $y = du_1/dK_R|_{K_R=K_C}, z = du_2/dK_R|_{K_R=K_C}, w = v_1(K_C, 0), x = v_2(K_C, 0)$ are constants.

Solving the first of these equations gives

$$K_R = K_C + (K - K_C)R^z f(R)^y,$$

(3.5)

again remembering that $K_1 = K, J_1 = J, f(1) = 1$.

Wilson's essential trick is to iterate this relation, which is supposed to converge rapidly near the critical point (Binney et al., 1986), assuming that for K_R near K_C, we have

$$K_C/2 \approx K_C + (K - K_C)R^z f(R)^y.$$

(3.6)

We iterate in two steps, first solving this for $f(R)$ in terms of known values, and then solving for R, finding a value R_C that we then substitute into the first of equations (3.1) to obtain an expression for $H[K,0]$ in terms of known functions and parameter values.

The first step gives the general result

$$f(R_C) \approx \frac{[K_C/(K_C - K)]^{1/y}}{2^{1/y}R_C^{z/y}}.$$

(3.7)

Solving this for R_C and substituting into the first expression of equation (3.1) gives, as a first iteration of a far more general procedure (Shirkov and Kovalev, 2001), the result

$$H[K,0] \approx \frac{H[K_C/2,0]}{f(R_C)} = \frac{H_0}{f(R_C)}$$

$$\chi(K,0) \approx \chi(K_C/2,0)R_C = \chi_0 R_C,$$

(3.8)

which are the essential relationships.

Note that a power law of the form $f(R) = R^m, m = 3$, which is the direct physical analog, may not be biologically reasonable, since it says that 'language richness' can grow very rapidly as a function of increased network size. Such rapid growth is simply not observed.

Taking the biologically realistic example of non-integral 'fractal' exponential growth,

$$f(R) = R^\delta,$$

(3.9)

where $\delta > 0$ is a real number which may be quite small, equation (3.7) can be solved for R_C, obtaining

$$R_C = \frac{[K_C/(K_C - K)]^{[1/(\delta y + z)]}}{2^{1/(\delta y + z)}}$$

(3.10)

for K near K_C. Note that, for a given value of y, one might characterize the relation $\alpha \equiv \delta y + z = $ constant as a 'tunable universality class relation' in the sense of Albert and Barabasi (2002).

Substituting this value for R_C back into equation (3.7) gives a complex expression for H, having three parameters: δ, y, z.

A more biologically interesting choice for $f(R)$ is a logarithmic curve that 'tops out', for example

$$f(R) = m \log(R) + 1.$$

(3.11)

Again $f(1) = 1$.

Using Mathematica 4.2 or above to solve equation (3.7) for R_C gives

$$R_C = [\frac{Q}{W[Q \exp(z/my)]}]^{y/z},$$

(3.12)

where

$$Q \equiv (z/my)2^{-1/y}[K_C/(K_C - K)]^{1/y}.$$

The transcendental function $W(x)$, the Lambert W-function, is defined by the relation

$$W(x) \exp(W(x)) = x.$$

It arises in the theory of random networks and in renormalization strategies for quantum field theories.

An asymptotic relation for $f(R)$ would be of particular biological interest, implying that 'language richness' increases to a limiting value with population growth. Such a pattern is broadly consistent with calculations of the degree of allelic heterozygosity as a function of population size under a balance between genetic drift and neutral mutation (Hartl and Clark, 1997; Ridley, 1996). Taking

$$f(R) = \exp[m(R-1)/R]$$

(3.13)

gives a system which begins at 1 when $R = 1$, and approaches the asymptotic limit $\exp(m)$ as $R \to \infty$. Mathematica 4.2 finds

$$R_C = \frac{my/z}{W[A]},$$

(3.14)

where

$$A \equiv (my/z) \exp(my/z)[2^{1/y}[K_C/(K_C - K)]^{-1/y}]^{y/z}.$$

These developments indicate the possibility of taking the theory significantly beyond arguments by abduction from simple physical models, although the notorious difficulty of implementing information theory existence arguments will undoubtedly persist.

3.3 Universality class distribution

Physical systems undergoing phase transition usually have relatively pure renormalization properties, with quite different systems clumped into the same 'universality class,' having fixed exponents at transition (Binney et al., 1986). Biological and social phenomena may be far more complicated.

If the system of interest is a mix of subgroups with different values of some significant renormalization parameter m in the expression for $f(R, m)$, according to a distribution $\rho(m)$, then the first expression in equation (3.1) should generalize, at least to first order, as

$$H[K_R, J_R] = < f(R, m) > H[K, J]$$

$$\equiv H[K, J] \int f(R, m)\rho(m)dm.$$

(3.15)

If $f(R) = 1 + m \log(R)$ then, given any distribution for m,

$$< f(R) >= 1+ < m > \log(R),$$

(3.16)

where $< m >$ is simply the mean of m over that distribution.

Other forms of $f(R)$ having more complicated dependencies on the distributed parameter or parameters, like the power law R^δ, do not produce such a simple result. Taking $\rho(\delta)$ as a normal distribution, for example, gives

$$< R^\delta >= R^{<\delta>} \exp[(1/2)(\log(R^\sigma))^2],$$

(3.17)

where σ^2 is the distribution variance. The renormalization properties of this function can be determined from equation (3.7), and the calculation is left to the reader as an exercise, best done in Mathematica 4.2 or above.

Thus the information dynamic phase transition properties of mixed systems will not in general be simply related to those of a single subcomponent, a matter of possible empirical importance. If sets of relevant parameters defining renormalization universality classes are indeed distributed, experiments observing pure phase changes may be very difficult. Tuning among different possible renormalization strategies in response to external signals would result in even greater ambiguity in recognizing and classifying information dynamic phase transitions.

Important aspects of mechanism may be reflected in the combination of renormalization properties and the details of their distribution across subsystems.

In sum, real biological, social, or interacting biopsychosocial systems are likely to have very rich patterns of phase transition which may not display the simplistic, indeed, literally elemental, purity familiar to physicists. Overall mechanisms will, however, still remain significantly constrained by the theory, in the general sense of probability limit theorems.

3.4 Punctuated universality class tuning

The next step is to iterate the general argument onto the process of phase transition itself, producing a tunable workspace model subject to inherent punctuated detection of external events.

As described above, an essential character of physical systems subject to phase transition is that they belong to particular 'universality classes'. Again, this means that the exponents of power laws describing behavior at phase transition will be the same for large groups of markedly different systems, with 'natural' aggregations representing fundamental class properties (Binney et al., 1986).

It appears that biological networks undergoing phase transition analogs need not be constrained to such classes, and that 'universality class tuning', meaning the strategic alteration of parameters characterizing the renormalization properties of punctuation, might well be possible. Here we focus on the tuning of parameters within a single given renormalization relation. Clearly, however, wholesale shifts of renormalization properties must ultimately be considered as well, a matter for future work.

Universality class tuning has been observed in models of 'real world' networks. As Albert and Barabasi (2002) put it,

"The inseparability of the topology and dynamics of evolving networks is shown by the fact that [the exponents defining universality class] are related by [a] scaling relation..., underlying the fact that a network's assembly uniquely determines its topology. However, in no case are these exponents unique. They can be tuned continuously..."

Suppose that a structured external environment, itself an appropriately regular information source **Y**, 'engages' a modifiable cognitive system. The environment begins to write an image of itself on the cognitive system in a distorted manner permitting definition of a mutual information $I[K]$ splitting criterion according to the Rate Distortion or Joint Asymptotic Equipartition Theorems. K is an inverse coupling parameter between system and environment. At punctuation – near some critical point K_C – the systems begin to interact very strongly indeed, and, near K_C, using the simple physical model of equation (3.2),

$$I[K] \approx I_0 [\frac{K_C - K}{K_C}]^\alpha.$$

For a physical system α is fixed, determined by the underlying 'universality class.' Here we will allow α to vary, and, in the section below, to itself respond explicitly to signals.

Normalizing K_C and I_0 to 1,

$$I[K] \approx (1 - K)^\alpha.$$

(3.18)

The horizontal line $I[K] = 1$ corresponds to $\alpha = 0$, while $\alpha = 1$ gives a declining straight line with unit slope which passes through 0 at $K = 1$. Consideration shows there are progressively sharper transitions between the necessary zero value at $K = 1$ and the values defined by this relation for $0 < K, \alpha < 1$. The rapidly rising slope of transition with declining α is of considerable significance:

The instability associated with the splitting criterion $I[K]$ is defined by

$$Q[K] \equiv -KdI[K]/dK = \alpha K(1 - K)^{\alpha-1},$$

(3.19)

and is singular at $K = K_C = 1$ for $0 < \alpha < 1$. Following the arguments of section 2.8, we interpret this to mean that values of $0 < \alpha \ll 1$ are highly unlikely for real systems, since $Q[K]$, in this model, represents a kind of barrier for 'social' information systems, in particular interacting network modules.

On the other hand, smaller values of α mean that the system is far more efficient at responding to the adaptive demands imposed by the embedding structured environment, since the mutual information which tracks the matching of internal response to external demands, $I[K]$, rises more and more quickly toward the maximum for smaller and smaller α as the inverse coupling parameter K declines below $K_C = 1$. That is, in this model, systems able to attain smaller α are more responsive to external signals than those characterized by larger values. But smaller values will be harder to reach, probably only at some considerable physiological or opportunity cost. Focused cognitive action takes resources, of one form or another.

Wallace (2005a) makes these considerations explicit, modeling the role of contextual and energy constraints on the relations between Q, I, and other system properties.

The more biologically realistic renormalization strategies given above produce sets of several parameters defining the universality class, whose tuning gives behavior much like that of α in this simple example.

Formal iteration of the phase transition argument on this calculation gives tunable higher level cognition, focusing on paths of universality class parameters.

Suppose the renormalization properties of a language-on-a-network system at some 'time' k are characterized by a set of parameters $A_k \equiv \alpha_1^k, ..., \alpha_m^k$. Fixed parameter values define a particular universality class for the renormalization. We suppose that, over a sequence of 'times,' the universality class properties can be characterized by a path $x_n = A_0, A_1, ..., A_{n-1}$ having significant serial correlations. The correlations, in fact, permit definition of an adiabatically piecewise stationary ergodic information source associated with the paths x_n. We call that source $\mathbf{X}$.

Suppose also, in the now-usual manner, that the set of impinging external (or internal, systemic) signals is also highly structured and forms another information source $\mathbf{Y}$ which interacts not only with the system of interest globally, but specifically with its universality class properties as characterized by $\mathbf{X}$. $\mathbf{Y}$ is necessarily associated with a set of paths y_n.

Pair the two sets of paths into a joint path $z_n \equiv (x_n, y_y)$ and invoke an inverse coupling parameter, K, between the information sources and their paths. This leads, by the arguments above, to phase transition punctuation of $I[K]$, the mutual information between $\mathbf{X}$ and $\mathbf{Y}$, under either the Joint Asymptotic Equipartition Theorem or under limitation by a distortion measure, through the Rate Distortion Theorem. The essential point is that $I[K]$ is a splitting criterion under these theorems, and thus partakes of the homology with free energy density which we have invoked above.

Activation of universality class tuning, the mean field model's version of attentional focusing, then becomes itself a punctuated event in response to increasing linkage between the organism and an external structured signal or some particular system of internal events.

This iterated argument exactly parallels the extension of the General Linear Model to the Hierarchical Linear Model in regression theory (Byrk and Raudenbusch, 2001).

Another path to the fluctuating dynamic threshold might be through a second order iteration similar to that just above, but focused on the parameters defining the universality class distributions.

3.5 Another network of dynamic manifolds and its tuning

The set of universality class tuning parameters, A_k, defines another topological manifold in the sense of section 2.8, whose topology could also be more fully analyzed using Morse theory. That is, the arguments leading to section 2.9 can be applied here as well: the equivalence class of dynamic manifolds is determined, not by universality class, which is tunable, but by the underlying form of the renormalization relation, in the sense of the many different possible renormalization symmetries described in section 3.2, equations 3.9, 3.11 and 3.13, etc. Thus the possible higher level dynamic manifolds in this model are characterized by fixed renormalization relations, but tunable universality class parameters. In precisely the sense of section 2.9 one can invoke a crosstalk coupling within a groupoid network of different dynamic manifolds defined by these renormalization relations, leading to the same kind of Morse theoretic analysis of the higher level topological structure.

3.6 Evolutionary implications of multiple models

The existence of two distinct analytically solvable models for cognitive gene expression, the mean number versions of chapter 2 and the mean field forms of this chapter, suggests the possibility that evolution may have explored a broad spectrum of less mathematically tractable approaches to the phenomenon.

If evolutionary history is any indication, current patterns of cognitive gene expression are unlikely plesiomorphic. Evolution is littered with innovative flux. Multicellularity began among the Ediacara 650 million years ago and arose again in the Tommotian phase of the Cambrian Explosion, with a diversity of body plans unrivaled since. The Tetrapoda's late Devonian land invasion included more than one multiple-digit plan. *Ichthyostega, Acanthostega*, and *Tulerpeton* emerged with six, seven, even eight digits on each limb. Varieties of great apes far outnumbering present-day remnants brachiated through the Miocene.

We hypothesize similar evolutionary radiations of cognitive gene expression, wherein bouts of divergent innovations arose together. In all likelihood multiple attempts were made at building working gene expression mechanisms.

The key point here is that plentiful data indicate evolution has long experimented with the means and modes by which cognitive modules are interconnected. The resulting referential experiences the organisms partake are likely quite different even as the underlying mechanisms are phylogenetically related.

In our scenario, the resulting phylogenetic bushes of cognitive gene expression were subsequently pruned by some combination of blind chance and failure to adapt to changing circumstances. If true, we are left with little recourse in excavating the resulting fossils, although comparison of gene expression mechanisms across taxa may prove of interest in reconstructing the evolutionary history of the processes involved. Much effort has been made to trace how deep through the Tree of Life molecular homologies run (e.g., Hox genes). But homoplasies – by convergence and by horizontal capture – litter the genome too. The vertebrate immunoglobin super family, for one, including alternate transposition by RAG1 and RAG2 (Agrawal et al., 1998), appears viral in origin. Innovations of polyphyletic origin are routinely borrowed, making organisms, although related by monophyletic descent, rough mosaics.

Next we develop a hybrid model of highly punctuated coevolutionary interaction from these perspectives, relating ecosystem, gene expression, and evolutionary dynamics in a single structure dominated by mesoscale processes of ecological resilience.

4

Coevolution

4.1 The basic idea

Natural systems subject to coevolutionary interaction may become enmeshed in the Red Queen dilemma of Alice in Wonderland, in that they must undergo constant evolutionary change in order to avoid extirpation (Van Valen, 1973). They must constantly run just to stay in the same place. An example would be a competitive arms race between predator and prey. Each evolutionary advance in predation must be met with a coevolutionary adaptation which allows the prey to avoid the more efficient predator. Otherwise the system cannot persist, since a highly specialized predator can literally eat itself to extinction. Similarly, each prey defense must be matched by predator adaptation.

Here we present a fairly elaborate model of coevolution, in terms of interacting information sources. Interaction events, we will argue, can be highly punctuated. These may be between Darwinian genetic, cognitive, or embedding ecosystem structures.

4.2 Fragmentation and coalescence

We begin by reexamining ergodic information sources and their dynamics under the self-similarity of a renormalization transformation near a punctuated phase transition. We then study the linked interaction of two information sources in which the richness of the quasi-language of each affects the other, that is, when two information sources have become one another's primary environments. This leads directly and naturally to a coevolutionary Red Queen. We will generalize the development to a 'block diagonal' structure of several interacting sources, and extend the model to include the influence of embedding 'cultural' and other contexts, producing a 'farming' framework.

The structures of interest to us here can be most weakly, and hence universally, described in terms of an adiabatically, piecewise stationary, ergodic information source involving a stochastic variate X. In some general sense,

R. Wallace et al., *Farming Human Pathogens*, DOI 10.1007/978-0-387-92213-3_4,

the source sends symbols α in correlated sequences $\alpha_0, \alpha_1 ... \alpha_{n-1}$ of length n (which may vary), according to a joint probability distribution, and its associated conditional probability distribution,

$$P[X_0 = \alpha_0, X_1 = \alpha_1, ... X_{n-1} = \alpha_{n-1}],$$

$$P[X_{n-1} = \alpha_{n-1} | X_0 = \alpha_0, ... X_{n-2} = \alpha_{n-2}].$$

If the conditional probability distribution depends only on m previous values of X, then the information source is said to be of order m (Ash, 1990).

Again, by 'ergodic' we mean that, in the long term, correlated sequences of symbols are generated at an average rate equal to their (joint) probabilities. 'Adiabatic' means that changes are slow enough to allow the necessary limit theorems to function. 'Stationary' means that, between pieces, probabilities don't change (much), and 'piecewise' means that these properties hold between phase transitions, which are described using renormalization methods.

As the length of the correlated sequences increases without limit, the Shannon-McMillan Theorem permits division of all possible streams of symbols into two groups, a relatively small number characterized as meaningful, whose long-time behavior matches the underlying probability distribution, and an increasingly large set of gibberish with vanishingly small probability. Let $N(n)$ be the number of possible meaningful sequences of length n emitted by the source $\mathbf{X}$. Again, uncertainty of the source, $H[\mathbf{X}]$, can be defined as

$$H[\mathbf{X}] = \lim_{n \to \infty} \frac{\log[N(n)]}{n}.$$

The Shannon-McMillan Theorem shows how to characterize $H[\mathbf{X}]$ directly in terms of the joint probability distribution of the source $\mathbf{X}$: $H[\mathbf{X}]$ is observable and can be calculated from the inferred pattern of joint probabilities.

Let $P[x_i | y_j]$ be the conditional probability that stochastic variate $X = x_i$ given that stochastic variate $Y = y_j$ and let $P[x_i, y_j]$ be the joint probability that $X = x_i$ and $Y = y_j$. Then the joint and conditional uncertainties of X and Y, $H(X, Y)$, and $H(X|Y)$ are given by expressions like those of equation (1.4).

And again, the Shannon-McMillan Theorem of states that $H[\mathbf{X}]$ is given by the limits of equation (1.5).

Estimating the probabilities of the sequences $\alpha_0, ... \alpha_{n-1}$ from observation, the ergodic property allows us to use them to estimate the uncertainty of the source, of the behavioral language $\mathbf{X}$. That is, $H[\mathbf{X}]$ is directly measurable.

Some elementary consideration (Ash, 1990; Cover and Thomas, 1991) shows that source uncertainty has a least upper bound, a supremum, defined by the capacity of the channel along which information is transmitted. That is, there exists a number C defined by externalities such that $H[\mathbf{X}] \leq C$.

C is the maximum rate at which the external world can transmit information originating with the information source, or that internal workspaces can communicate. Much of the subsequent development could, in fact, be expressed using this inequality, which will be generalized in the following chapter.

Again recall the relation between the expression for source uncertainty and the free energy density of a physical system, as expressed by equation (2.4), which undergoes a phase transition depending on an inverse temperature parameter $K = 1/T$ at a critical temperature T_C.

Imposition of a renormalization symmetry on $F(K)$ in equation (2.4) describes, in the infinite volume limit, the behavior of the system at the phase transition in terms of scaling laws (K. Wilson, 1971). After some development, taking the limit $n \to \infty$ as an analog to the infinite volume limit of a physical system, we will apply this approach to a parametized source uncertainty. We will examine changes in structure as a fundamental 'inverse temperature' changes across the underlying system.

We use three parameters to describe the relations between an information source and its environment or between different interacting sources. The first, $J \geq 0$, measures the degree to which acquired characteristics are transmitted. $J \approx 0$ thus represents a high degree of genetic as opposed to cultural inheritance. J will always remain distinguished, a kind of inherent direction or external field strength in the sense of Wilson (1971).

The second parameter, $Q = 1/\mathcal{C} \geq 0$, represents the inverse availability of resources. $Q \approx 0$ thus represents a high ability to renew and maintain a system.

The third parameter, $K = 1/T$, is an inverse index of a generalized temperature T, which we will more directly specify below in terms of the richness of interacting information sources.

We suppose further that the structure of interest is implicitly embedded in, and operates within the context of, a larger manifold stratified by metric distances.

Take these as multidimensional vector quantities $\mathbf{A}$, $\mathbf{B}$, $\mathbf{C}$.... $\mathbf{A}$ may represent location in space, time delay, or the like, and $\mathbf{B}$ may be determined through multivariate analysis of a spectrum of observed behavioral or other factors, in the largest sense, etc.

It may be possible to reduce the effects of these vectors to a function of their magnitudes $a = |\mathbf{A}|$, $b = |\mathbf{B}|$ and $c = |\mathbf{C}|$, etc. Define the *generalized distance* r as

$$r^2 = a^2 + b^2 + c^2 + ...$$

(4.1)

To be more explicit, we assume an ergodic information source **X** is associated with the reproduction and/or persistence of a population, ecosystem, cognitive dual language or other structure. The source **X**, its uncertainty $H[J, K, Q, \mathbf{X}]$, and its parameters J, K, Q all are assumed to depend *implicitly* on the embedding manifold, in particular on the metric r of equation (4.1).

A particularly elegant and natural formalism for generating such punctuation in our context involves application of Wilson's (1971) program of renormalization symmetry – invariance under the renormalization transform – to source uncertainty defined on the r-manifold. The results predict that language in the most general sense, which includes the transfer of information within a a cognitive enterprise, or between an enterprise and an embedding context, will undergo sudden changes in structure analogous to phase transitions in physical systems.

We must, however, emphasize that this approach is argument by abduction, in Hodgson's (1993) sense, from physical theory. Much current development surrounding self-organizing physical phenomena is based on the assumption that at phase transition a system looks the same under renormalization. That is, phase transition represents a stationary point for a renormalization transform in the sense that the transformed quantities are related by simple scaling laws to the original values.

Renormalization is a clustering semigroup transformation in which individual components of a system are combined according to a particular set of rules into a 'clumped' system whose behavior is a simplified average of those components. Since such clumping is a many-to-one condensation, there can be no unique inverse renormalization, and, as Chapter 3 shows, many possible forms of condensation.

Assume it possible to redefine characteristics of the information source **X** and J, K, Q as functions of averages across the manifold having metric r, which we write as R. That is, 'renormalize' by clustering the entire system in terms of blocks of different sized R.

Let $N(K, J, Q, n)$ be the number of high probability meaningful correlated sequences of length n *across the entire community* in the r-manifold of equation (4.1), given parameter values K, J, Q. We study changes in

$$H[K, J, Q, \mathbf{X}] \equiv \lim_{n \to \infty} \frac{\log[N(K, J, Q, n)]}{n}$$

as $K \to K_C$ and/or $Q \to Q_C$ for critical values K_C, Q_C at which the system begins to undergo a marked transformation from one kind of structure to another.

Given the metric of equation (4.1), a *correlation length*, $\chi(K, J, Q)$, can be defined as the average length in r-space over which structures involving a particular phase dominate.

Now clump the 'community' into blocks of average size R in the multivariate r-manifold, the 'space' in which the system of interest is implicitly embedded.

Following the arguments of Chapter 3 it is possible to impose renormalization symmetry on the source uncertainty on H and χ by assuming at transition the relations

$$H[K_R, J_R, Q_R, \mathbf{X}] = R^\delta H[K, J, Q, \mathbf{X}]$$

(4.2)

and

$$\chi(K_R, J_R, Q_R) = \frac{\chi(K, J, Q)}{R}$$

(4.3)

hold, where K_R, J_R and Q_R are the transformed values of K, J and Q after the clumping of renormalization. We take $K_1, J_1, Q_1 \equiv K, J, Q$ and permit the characteristic exponent $\delta > 0$ to be real. Chapter 3 provides examples of other possible relations. Equations (4.2) and (4.3) are assumed to hold in a neighborhood of the transition values K_C and Q_C.

Differentiating these with respect to R gives complicated expressions for dK_R/dR, dJ_R/dR and dQ_R/dR depending simply on R which we write as

$$dK_R/dR = \frac{u(K_R, J_R, Q_R)}{R}$$

$$dQ_R/dR = \frac{w(K_R, J_R, Q_R)}{R}$$

$$dJ_R/dR = \frac{v(K_R, J_R, Q_R)}{R} J_R.$$

(4.4)

Solving these differential equations gives K_R, J_R and Q_R as functions of J, K, Q and R.

Substituting back into equations (4.2) and (4.3) and expanding in a first order Taylor series near the critical values K_C and Q_C gives power laws much like the Widom-Kadanoff relations for physical systems (Wilson, 1971). For example, letting $J, Q \to 0$ and taking $\kappa \equiv (K_C - K)/K_C$ gives, in first order near K_C,

$$H = \kappa^{\delta/y} H_0$$

$$\chi = \kappa^{-1/y} \chi_0$$

(4.5)

where y is a constant arising from the series expansion.

Note that there are only two fundamental equations – (4.2) and (4.3) – in $n > 2$ unknowns: The critical 'point' is, in this formulation, most likely to be a complicated implicitly defined critical surface in $J, K, Q, ...$-space. The 'external field strength' J remains distinguished in this treatment – the inverse of the degree to which acquired characteristics are inherited – but *neither K, Q nor other parameters are, by themselves, fundamental*, rather their joint interaction defines critical behavior along this surface.

The surface is a fundamental object, not the particular set of parameters (except for J) used to define it. That surface may be subject to any set of transformations leaving the surface invariant. Thus inverse generalized temperature resource availability or whatever other parameters may be identified as affecting the richness of cognition, are inextricably intertwined and mutually interacting, according to the form of this critical evolutionary transition surface. That surface, in turn, is unlikely to remain fixed, and should vary with time or other extrinsic parameters, including, but not likely limited to, J.

At the critical surface a Taylor expansion of the renormalization equations (4.2) and (4.3) gives a first order matrix of derivatives whose eigenstructure defines fundamental system behavior. For physical systems the surface is a saddle point (Wilson, 1971), but more complicated behavior seems likely in the ecological and social systems we study. See Binney et al., (1986) for some details of this differential geometry.

Taking, for the moment, the simplest formulation, $(J, Q \to 0)$, as K increases toward a threshold value K_C, the source uncertainty of the reproductive, behavioral, or other language common across the system declines and, at K_C, the average regime dominated by the 'other phase' grows. That is, the system begins to freeze into one having a large correlation length for the second phase. The two phenomena are linked at criticality in physical systems by the scaling exponent y.

Assume the rate of change of $\kappa = (K_C - K)/K_C$ remains constant, $|d\kappa/dt| = 1/\tau_K$. Analogs with physical theory suggest there is a characteristic time constant for the phase transition. $\tau \equiv \tau_0/\kappa$, such that if changes in κ take place on a timescale longer than τ for any given κ, we may expect the correlation length $\chi = \chi_0 \kappa^{-s}$, $s = 1/y$, will be in equilibrium with internal changes and result in a very large fragment in r-space. Following Zurek (1985, 1996), the 'critical' freezeout time, $\hat{t}$, will occur at a 'system time' $\hat{t} = \chi/|d\chi/dt|$ such that $\hat{t} = \tau$. Taking the derivative $d\chi/dt$, remembering that by definition $d\kappa/dt = 1/\tau_K$, gives

$$\frac{\chi}{|d\chi/dt|} = \frac{\kappa \tau_K}{s} = \frac{\tau_0}{\kappa},$$

so that

$$\kappa = \sqrt{s\tau_0/\tau_K}.$$

Substituting this value of κ into the equation for correlation length, the expected size of fragments in r-space, $d(\hat{t})$, becomes

$$d \approx \chi_0 (\frac{\tau_K}{s\tau_0})^{s/2},$$

with $s = 1/y > 0$. The more rapidly K approaches K_C the smaller is τ_K and the smaller and more numerous are the resulting r-space fragments. Thus rapid change produces small fragments more likely to risk extinction in a system dominated by economies of scale.

It is very important to recognize that this process can proceed in the opposite direction, i.e., punctuated coalescence from a collection of relatively independent, loosely interacting and largely isolated subcomponents, into a far more coherent and integrated structure having very complicated dynamics of mutual influence.

The next section focuses specifically on the analysis of those dynamics.

4.3 Recursive interaction

Extending the theory involves envisioning reciprocally interacting genetic, cognitive, or ecosystem information sources as subject to a coevolutionary Red Queen in the sense of Van Valen (1973) by treating their respective source uncertainties as recursively parametized by each other. That is, *assume the information sources are each other's primary environments.* These are, respectively, characterized by information sources **X** and **Y**, whose uncertainties are parametized

[1] by measures of both inheritance and inverse resources – $J\,Q$ as above – and, most critically,

[2] by each others inverse uncertainties, $\mathcal{H}_X \equiv 1/H[\mathbf{X}]$ and $\mathcal{H}_Y \equiv 1/H[\mathbf{Y}]$, so that

$$H[\mathbf{X}] = H[Q, J, \mathcal{H}_Y, \mathbf{X}]$$

$$H[\mathbf{Y}] = H[Q, J, \mathcal{H}_X, \mathbf{Y}].$$

(4.6)

This is a recursive system having complex behaviors.

Assume a strongly heritable genetic system, $J \to 0$, with fixed inverse resource base, Q, for which $H[\mathbf{X}]$ follows something like the lower graph in figure 4.1, a reverse S-shaped curve with $K \equiv \mathcal{H}_Y = 1/H[\mathbf{Y}]$, and similarly $H[\mathbf{Y}]$ depends on $\mathcal{H}_X$. That is, increase or decline in the source uncertainty of one system leads to increase or decline in the source uncertainty of the other. The richness of the two information sources is closely linked.

Start at the right of the lower graph for $H[\mathbf{X}]$ in figure 4.1, the source uncertainty of the first system, but to the left of the critical point K_C. Assume $H[\mathbf{Y}]$ increases so $\mathcal{H}_Y$ decreases, and thus $H[\mathbf{X}]$ increases, walking up the lower curve of the figure from the right. The richness of the first system's internal language increases or the interaction between internal structures increases the richness of their dual cognitive information sources. They get smarter or faster or more poisonous, or their herd behavior becomes more sophisticated in the presence of a predator.

The increase of $H[\mathbf{X}]$ leads, in turn, to a decline in $\mathcal{H}_X$ and triggers an increase of $H[\mathbf{Y}]$, whose increase leads to a further increase of $H[\mathbf{X}]$ and vice versa. The Red Queen takes the system from the right of figure 4.1 to the left, up the lower curve as the two systems mutually interact.

Now enlarge the scale of the argument, and consider the possibility of other interactions.

The upper graph of figure 4.1 represents the disorder

$$S = H[K, \mathbf{X}] - KdH[K, \mathbf{X}]/dK, K \equiv 1/H[\mathbf{Y}].$$

According to the dynamical manifold analysis, the peak in S represents a repulsive barrier for transition between high and low values of $H[\mathbf{X}]$. This

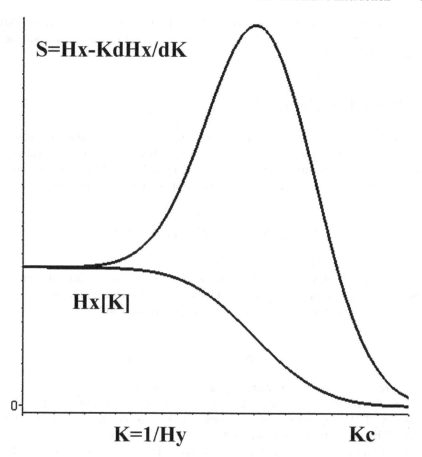

Fig. 4.1. A reverse-S-shaped curve for source uncertainty $H[\mathbf{X}]$ – measuring language richness – as a function of an inverse temperature parameter $K = 1/H[\mathbf{Y}]$. To the right of the critical point K_C the system breaks into fragments in r-space whose size is determined by the rate at which K approaches K_C. A collection of fragments already to the right of K_C, however, would be seen as condensing into a single unit as K declined below the critical point. If K is an inverse source uncertainty itself, i.e., $K = 1/H[\mathbf{Y}]$ for some information source $\mathbf{Y}$, then under such conditions a Red Queen dynamic can become enabled, driving the system strongly to the left. No intermediate points are asymptotically stable, given a genetic heritage in this development, although generalized Onsager/dynamical arguments suggest that the repulsive peak in $S = H - K/dH/dK$ can serve to create quasi-stable resilience realms. To the right of the critical point K_C the system is locked into disjoint fragments.

leads to the expectation of *hysteresis*. That is, the two realms, to the left and right of the peak in S for figure 4.1, thus represent quasi-stable resilience modes, in this model.

4.4 Extending the model

The model directly generalizes to multiple interacting information sources.

First consider a matrix of crosstalk measures between a set of information sources. Assume the matrix can be block diagonalized into two major components, characterized by network information source measures like equation (2.3),

$$I_m(X_1...X_i|Y_1...Y_j|Z_1...Z_k), m = 1, 2.$$

Then apply the two-component theory above.

Extending the development to multiple, recursively interacting information sources resulting from a more general block diagonalization seems direct. First use inverse measures $\mathcal{I}_j \equiv 1/I_j, j \neq m$ as parameters for each of the other blocks, writing

$$I_m = I_m(K_1...K_s, ...\mathcal{I}_j...), j \neq m,$$

where the K_s represent other relevant parameters.

Next segregate the $\mathcal{I}_j$ according to their relative rates of change, as in equation (2.3). Cognitive gene expression would be among the most rapid, followed by ecosystem dynamics and evolutionary selection.

The dynamics of such a system, following the pattern of equations (2.8) and (2.14), becomes a recursive network of stochastic differential equations, similar to those used to study many other highly parallel dynamic structures (Wymer, 1997).

Letting the K_j and $\mathcal{I}_m$ all be represented as parameters Q_j, (with the caveat that I_m not depend on $\mathcal{I}_m$), one can define

$$S_I^m \equiv I_m - \sum_i Q_i \partial I_m / \partial Q_i$$

to obtain a complicated recursive system of phenomenological 'Onsager relations' stochastic differential equations like (2.14),

$$dQ_t^j = \sum_i [L_{j,i}(t, ...\partial S_I^m / \partial Q^i ...)dt + \sigma_{j,i}(t, ...\partial S_I^m / \partial Q^i ...)dB_t^i]$$

$$= L_j(Q^1, ..., Q^n)dt + \sum_i \sigma(t, Q^1, .., Q^n)dB_t^i,$$

(4.7)

where we have collected terms and expressed both the reciprocal $\mathcal{I}$'s and the external K's in terms of the same Q_j. The second mathematical appendix provides an introduction to stochastic differential equations.

The index m ranges over the crosstalk and we could allow different kinds of 'noise' dB_t^i, having particular forms of quadratic variation which may, in fact, represent a projection of environmental factors under something like a rate distortion manifold.

Indeed, the I_m and/or the derived S^m might, in some circumstances, be combined into a Morse function, permitting application of Pettini's topological hypothesis.

The model rapidly becomes unwieldy, probably requiring some clever combinatorial or groupoid convolution algebra and related diagrams for concise expression, much as in the usual field theoretic perturbation expansions (Hopf algebras, for example). The virtual reaction method of Zhu et al. (2007) is another possible approach.

As in the simple model above, there will be, first, multiple quasi-stable points within a given system's I_m, representing a class of generalized resilience modes accessible via punctuation and enmeshing gene selection, gene expression, and ecological resilience – similar to figure 4.1.

Second, however, will be analogs to the fragmentation of figure 4.1 when the system exceeds the critical value K_c. That is, full-scale disintegration of the entire system, and not just punctuation within it.

We thus infer two classes of punctuation possible for this kind of structure, both of which could entrain ecosystem resilience shifts, gene expression, and gene selection, although the latter kind would seem to be the far more dramatic.

There are other possible patterns:

[1] Setting equation (4.7) equal to zero and solving for stationary points again gives attractor states since the noise terms preclude unstable equilibria.

[2] Unlike equation (2.14), however, this system may converge to limit cycle or pseudorandom 'strange attractor' behaviors in which the system seems to chase its tail endlessly within a limited venue – the traditional Red Queen.

[3] What is converged to in both cases is not a simple state or limit cycle of states. Rather it is an equivalence class, or set of them, of highly dynamic information sources coupled by mutual interaction through crosstalk. Thus 'stability' in this structure represents particular patterns of ongoing dynamics rather than some identifiable static configuration.

Here we are, at last and indeed, deeply enmeshed in a highly recursive phenomenological stochastic differential equations, but at a deeper level than

Zhu et al. (2007) envisioned for gene expression alone, and in a dynamic rather than static manner. The objects of this dynamical system are equivalence classes of information sources and their crosstalk, rather than simple 'states' of a dynamical or reactive chemical system.

Imposition of necessary conditions from the asymptotic limit theorems of communication theory has beaten the mathematical thicket back one full layer. Other formulations may well be possible, but our work here serves to illustrate the method.

It is, however, interesting to compare our results to those of Dieckmann and Law (1996), who invoke evolutionary game dynamics to obtain a first order canonical equation for coevolutionary systems having the form

$$ds_i/dt = K_i(s)\partial W_i(s_i', s)|_{s_i'=s_i}.$$

(4.8)

The s_i, with $i = 1, ..., N$ denote adaptive trait values in a community comprising N species. The $W_i(s_i', s)$ are measures of fitness of individuals with trait values s_i' in the environment determined by the resident trait values s, and the $K_i(s)$ are non-negative coefficients, possibly distinct for each species, that scale the rate of evolutionary change. Adaptive dynamics of this kind have frequently been postulated, based either on the notion of a hill-climbing process on an adaptive landscape or some other sort of plausibility argument.

When this equation is set equal to zero, so there is no time dependence, one obtains what are characterized as 'evolutionary singularities' or stationary points.

Dieckmann and Law contend that their formal derivation of this equation satisfies four critical requirements:

[1] The evolutionary process needs to be considered in a coevolutionary context.

[2] A proper mathematical theory of evolution should be dynamical.

[3] The coevolutionary dynamics ought to be underpinned by a microscopic theory.

[4] The evolutionary process has important stochastic elements.

Our equation (4.7) seems clearly within the same ballpark, although we have taken a much different route, one which produces elaborate patterns of phase transition punctuation in a highly natural manner. Champagnat et al. (2006), in fact, derive a higher order canonical approximation extending equation (4.8) which is very much closer equation to (4.7), that is, a stochastic differential equation describing evolutionary dynamics. Champagnat et al. (2006) go even further, using a large deviations argument to analyze dynamical coevolutionary paths, not merely evolutionary singularities. They contend that in general, the issue of evolutionary dynamics drifting away from trajectories predicted by the canonical equation can be investigated by considering the asymptotic of the probability of 'rare events' for the sample paths of the diffusion. By 'rare events' they mean diffusion paths drifting far away from the canonical equation. The probability of such rare events is governed by a large deviation principle: when a critical parameter (designated ϵ) goes to zero, the probability that the sample path of the diffusion is close to a given rare path ϕ decreases exponentially to 0 with rate $I(\phi)$, where the 'rate function' I can be expressed in terms of the parameters of the diffusion. This result, in their view, can be used to study long-time behavior of the diffusion process when there are multiple attractive evolutionary singularities. Under proper conditions the most likely path followed by the diffusion when exiting a basin of attraction is the one minimizing the rate function I over all the appropriate trajectories. The time needed to exit the basin is of the order $\exp(H/\epsilon)$ where H is a quasi-potential representing the minimum of the rate function I over all possible trajectories.

An essential fact of large deviations theory is that the rate function I which Champagnat et al. (2006) invoke can almost always be expressed as a kind of entropy, that is, in the form $I = -\sum_j P_j \log(P_j)$ for some probability distribution. This result goes under a number of names; Sanov's Theorem, Cramer's Theorem, the Gartner-Ellis Theorem, the Shannon-McMillan Theorem, and so forth (Dembo and Zeitouni, 1998). We suggest that gene expression, because of its underlying cognitive nature, may be an even more central aspect of coevolutionary process than is currently understood. The fluctuational paths defined by the system of equations in (4.7) may become serially correlated outputs of an information source driven by cognitive gene expression. In particular, the coevolutionary pressures inherent to equation (4.7) may in fact strongly select for significant cognition in gene expression, in essence another version of the Baldwin effect.

4.5 The large deviations formalism

It is of some interest to explicitly carry through the program suggested by Campagnat et al. (2006), and to compare it with the 'natural' generalization of equation 4.7. We begin with a recapitulation of large deviations and fluctuation formalism.

Information source uncertainty, according to the Shannon-McMillan Theorem, serves as a splitting criterion between high and low probability sequences (or pairs of them) and displays the fundamental characteristic of a growing body of work in applied probability often termed the Large Deviations Program, (LDP). This seeks to unite information theory, statistical mechanics, and the theory of fluctuations under a single umbrella.

Following Dembo and Zeitouni, (1998, p.2), let $X_1, X_2, ...X_n$ be a sequence of independent, standard Normal, real-valued random variables and let

$$S_n = \frac{1}{n}\sum_{j=1}^{n} X_j.$$

(4.9)

Since S_n is again a Normal random variable with zero mean and variance $1/n$, for all $\delta > 0$

$$\lim_{n\to\infty} P(|S_n| \geq \delta) = 0,$$

(4.10)

where P is the probability that the absolute value of S_n is greater or equal to δ. Some manipulation, however, gives

$$P(|S_n| \geq \delta) = 1 - \frac{1}{\sqrt{2\pi}}\int_{-\delta\sqrt{n}}^{\delta\sqrt{n}} \exp(-x^2/2)dx,$$

(4.11)

so that

$$\lim_{n\to\infty} \frac{\log P(|S_n| \geq \delta)}{n} = -\delta^2/2$$

(4.12)

This can be rewritten for large n as

$$P(|S_n| \geq \delta) \approx \exp(-n\delta^2/2).$$

(4.13)

That is, for large n, the probability of a large deviation in S_n follows something much like the asymptotic equipartition relation of the Shannon-McMillan Theorem, so that meaningful paths of length n all have approximately the same probability $P(n) \propto \exp(-nH[\mathbf{X}])$.

Questions about meaningful paths appear suddenly as formally isomorphic to the central argument of the LDP which encompasses statistical mechanics, fluctuation theory, and information theory into a single structure (Dembo and Zeitouni, 1998).

Again, the cardinal tenet of large deviation theory is that the rate function $-\delta^2/2$ can, under proper circumstances, be expressed as a mathematical entropy having the standard form

$$-\sum_k p_k \log p_k,$$

(4.14)

for some set of probabilities p_k.

Next we briefly recapitulate part of the standard treatment of large fluctuations (Onsager and Machlup, 1953; Fredlin and Wentzell, 1998).

The macroscopic behavior of a complicated physical system in time is assumed to be described by the phenomenological Onsager relations giving large-scale fluxes as

$$\sum_i C_{i,j} dK_j/dt = \partial S/\partial K_i,$$

(4.15)

where the $C_{i,j}$ are appropriate constants, S is the system entropy, and the K_i are the generalized coordinates which parametize the system's free energy.

Entropy is defined from free energy F by a Legendre transform – more of which follows below:

$$S \equiv F - \sum_j K_j \partial F / \partial K_j,$$

where the K_j are appropriate system parameters.

Neglecting volume problems for the moment, free energy can be defined from the system's partition function Z as

$$F(K) = \log[Z(K)].$$

The partition function Z, in turn, is defined from the system Hamiltonian – defining the energy states – as

$$Z(K) = \sum_j \exp[-KE_j],$$

where K is an inverse temperature or other parameter and the E_j are the energy states.

Inverting the Onsager relations gives

$$dK_i/dt = \sum_j L_{i,j} \partial S / \partial K_j = L_i(K_1, ..., K_m, t) \equiv L_i(K, t).$$

(4.16)

The terms $\partial S / \partial K_i$ are macroscopic driving forces dependent on the entropy gradient.

Let a white Brownian noise $\epsilon(t)$ perturb the system, so that

$$dK_i/dt = \sum_j L_{i,j} \partial S / \partial K_j + \epsilon(t)$$

$$= L_i(K, t) + \epsilon(t),$$

(4.17)

where the time averages of ϵ are $< \epsilon(t) >= 0$ and $< \epsilon(t)\epsilon(0) >= D\delta(t)$. $\delta(t)$ is the Dirac delta function, and we take K as a vector in the K_i.

Following Luchinsky (1997), if the probability that the system starts at some initial macroscopic parameter state K_0 at time $t = 0$ and gets to the state $K(t)$ at time t is $P(K, t)$, then a somewhat subtle development (e.g., Feller, 1971) gives the forward Fokker-Planck equation for P:

$$\partial P(K,t)/\partial t = -\nabla \cdot (L(K,t)P(K,t)) + (D/2)\nabla^2 P(K,t).$$

(4.18)

In the limit of weak noise intensity this can be solved using the WKB (i.e., the eikonal) approximation, as follows. Take

$$P(K,t) = z(K,t)\exp(-s(K,t)/D).$$

(4.19)

$z(K,t)$ is a prefactor and $s(K,t)$ is a classical action satisfying the Hamilton-Jacobi equation, which can be solved by integrating the Hamiltonian equations of motion. The equation reexpresses $P(K,t)$ in the usual parametized negative exponential format.

Let $p \equiv \nabla s$. Substituting and collecting terms of similar order in D gives

$$dK/dt = p + L,$$

$$dp/dt = -\partial L/\partial Kp$$

(4.20)

and

$$-\partial s/\partial t \equiv h(K,p,t) = pL(K,t) + \frac{p^2}{2},$$

(4.21)

with $h(K,t)$ the Hamiltonian for appropriate boundary conditions.

Again following Luchinsky (1997), these Hamiltonian equations have two different types of solution, depending on p. For $p = 0, dK/dt = L(K,t)$, describing the system in the absence of noise. We expect that with finite noise intensity the system will give rise to a distribution about this deterministic path. Solutions for which $p \neq 0$ correspond to *optimal paths* along which the system will move with overwhelming probability.

These results can, however, again be directly derived as a special case of a Large Deviation Principle based on generalized entropies mathematically similar to Shannon's uncertainty from information theory, bypassing the Hamiltonian formulation entirely.

4.6 Farming a coevolutionary system

For a cognitive system characterized by a dual information source, of course, there is no Hamiltonian, but the generalized entropy or splitting criterion treatment still works. The trick is to do with information source uncertainty what is done here with a Hamiltonians. We will show this follows from a simple extension of the system described by equation 4.7 above.

Here we are concerned not with a random Brownian distortion of simple physical systems, but, invoking cognitive gene expression, with a possibly complex behavioral structure, in the largest sense, composed of quasi-independent actors for which *meaningful/optimal paths have extremely structured serial correlation, amounting to a grammar and syntax, precisely the fact which allows definition of an information source* and enables the use of the very sparse equipartition of the Shannon-McMillan and Rate Distortion Theorems.

In sum, to again paraphrase Luchinsky (1997), large fluctuations, although infrequent, are fundamental in a broad range of processes, and it was recognized by Onsager and Machlup (1953) that insight into the problem could be gained from studying the distribution of fluctuational paths along which the system moves to a given state. This distribution is a fundamental characteristic of the fluctuational dynamics, and its understanding leads toward control of fluctuations. Fluctuational motion from the vicinity of a stable state may occur along different paths. For large fluctuations, the distribution of these paths peaks sharply along an optimal, most probable, path. In the theory of large fluctuations, the pattern of optimal paths plays a role similar to that of the phase portrait in nonlinear dynamics.

In this development, meaningful paths driven by cognitive or other phenomena having an associated information source can play something of the role of optimal paths in the theory of large fluctuations which Champagnat et al. (2006) have invoked, but without benefit of a Hamiltonian.

In particular, the dispersion of the spectrum of cognitive gene expression within a species, affecting the ability to adapt to changing ecological niches, then becomes central to the mitigation of selection pressures generated by coevolutionary dynamics. Too limited a response repertoire will cause a species to become fully entrained into high probability dynamical fluctuational paths leading to punctuated extirpation. A broad spectrum allows a species to ride out much more of the coevolutionary selection pressure.

A sufficiently broad repertoire of cognitive gene expression responses leads, however, to the necessity of a second order coevolution model in which the high probability fluctuational paths defined by the system of equations (4.7) *are, in fact, themselves the output of some information source.* This is a model closely analogous to the second order cognitive structures needed to explain animal consciousness (Wallace, 2005a). Intuitively, this transition to 'cognitive coevolution' would be particularly likely under the influence of a strong system of epigenetic inheritance, that is, an animal culture extending the niche spectrum offered by cognitive gene expression alone. Thus we can expand this development to one encompassing biocultural coevolution.

Most simply, and rather elegantly, such an extended argument just adds an embedding context to the basic set of information sources presented in Chapter 1. Ecosystem, cognitive gene expression, and the mechanisms of genetic inheritance would then interact with a fourth information source, defined by a generalized language associated with an encompassing (animal or human) culture, characterized by an information source. The version of equation 2.3 in section 4.3 then becomes fourfold:

$$I_m(X_1, ..., X_i|Y_1, ..., Y_j|Z_1, ..., Z_k|U),$$

(4.22)

where the X, Y, Z are as before and U represents the embedding cultural information source.

A considerable simplification results from not differentiating among the $X, Y,$ and Z, so that designating them all simply as Q_j, one obtains, using equation (2.2),

$$I(Q_1, ..., Q_n|U) = H(U) + \sum_{j=1}^{n} H(Q_j|U) - H(Q_1, ..., Q_n, U).$$

The system of equations 4.7 then becomes one step more complicated, as do the corresponding dynamics. Some clever reformulation using combinatorial algebras and the associated loop-structure arguments seems needed here.

The central point is that a one-step extension of the system of equation 4.7 permits incorporating the influence of an external cultural context in farming a coevolutionary system, that is, in cultivating evolutionary and ecosystem structures, including something of their trajectories.

The next chapter applies these perspectives to the coevolutionary interaction of prebiotic systems, focusing on the RNA error catastrophe paradox. The second chapter that follows examines highly adaptive human pathogens farmed by public policy, reductionist intervention, and socioeconomic structure. These mesoscale contexts drive the coevolutionary ecology of infection. The resulting evolutionarily transformed organisms may display multiple drug resistance, increased virulence associated with changes in life history strategy, and other adaptations which make their control and containment more difficult at both the individual and population levels.

5

Eigen's paradox

5.1 Introduction

Here we examine prebiotic evolution using the coevolutionary machinery of
the last chapter to explore a possible linkage between a fundamental ecosystem
resilience shift and an important change in evolutionary process. We follow
closely the recent paper by R. Wallace and R.G. Wallace (2008).

Applying the homology between information source uncertainty and free
energy density, under rate distortion constraints, the famous prebiotic error
catastrophe of Manfred Eigen emerges as the lowest energy state for simple
prebiotic systems without error correction. Invoking compartmentalization
– 'vesicles' – and using a Red Queen argument, suggests that information
crosstalk between two or more properly interacting structures can initiate a
coevolutionary dynamic having at least two quasi-stable states. The first is a
low energy realm near the error threshold, and, depending on available energy,
the second can approach zero error as a limit. A large deviations argument
produces jet-like global transitions which, over sufficient time, may enable
shifts between the many quasi-stable modes available to more complicated
structures. The transitions 'lock in' to some subset of the various possible
low error rate chemical systems, which become subject to development by
selection and stochastic trajectories.

Energy availability, according to the model, is thus a powerful necessary
condition for low error rate replication, suggesting that some fundamental
prebiotic ecosystem transformation entrained reproductive fidelity. That is, a
basic metabolic shift, such as a new chemical cycle, onset of predation, and/or
photofixation of energy, was needed to enable low error rate reproduction.

This work, then, supports speculation that our RNA/DNA world may
indeed be only the chance result of a very broad prebiotic evolutionary phenomenon. Processes *in vitro*, or *ex planeta*, might have other outcomes.

Manfred Eigen's (1971, 1996) evolution model is a landmark attempt at
coherently relating evolution, molecular biology, and information theory. That
work views selection as condensation in information space, and evolution as a

R. Wallace et al., *Farming Human Pathogens*, DOI 10.1007/978-0-387-92213-3_5,
© Springer Science+Business Media, LLC 2009

succession of phase transformations. Indeed, Eigen's quasispecies model, with its error catastrophe, corresponds exactly to a phase transition in a two dimensional Ising system (Leuthausser, 1986). The essence of Eigen's paradox is that the error catastrophe limits genome length in RNA precursor organisms to much less than observed in DNA organisms having error correcting enzymes, which, themselves, cannot be created in the absence of just such a long genome.

As Holmes (2005) put it,

> "To create more genetic complexity, it is therefore necessary to encode more information in longer genes by using a replication system with greater fidelity. But there's the catch: to replicate with greater fidelity requires a more accurate and hence complex replication enzyme, but such an enzyme cannot be created because this will itself require a longer gene, and longer genes will breach the error threshold".

Here we reconsider Eigen's paradox from a highly formal perspective which hews quite closely to the fundamental asymptotic limit theorems of information theory, in particular the Rate Distortion Theorem, and its zero-error limit, the Shannon-McMillan Theorem. These, like the Central Limit Theorem in parametric statistics, permit derivation of 'regression-like' models which can be applied to real data. We use these theorems in a principled manner to derive the high rate of mutation inherent to RNA virus replication and suggest a plausible Red Queen coevolutionary ratchet leading toward an evolutionary condensation resulting in effective error-correction mechanisms.

The line of argument is as follows:

[1] An increasingly complicated network of simple interacting 'RNA-like' organisms creates a collective biochemical system – a 'vesicle' – which, as a parallel communication channel, can have a much higher channel capacity for low-error replication than do the individual components.

[2] Several such distinct, properly interacting, collectives – compartments in the sense of Eigen and Szathmary – become each others' most intimate environments. The resulting coevolutionary ratchet produces a Red Queen structure which, given sufficient energy, can support quasi-stable states with very low reproductive error rates.

[3] High error rate, but low energy, systems-of-vesicles can become subject to systematic 'large deviations' excursions that, over sufficient time, can lead to the establishment of a distribution of low error rate, but higher energy, chemical systems. Even prebiological quasi-organisms can, apparently, build pyramids, as it were. Thus many different chemistry-of-life solutions seem possible, each subject to evolutionary processes of selection and chance extinction.

[4] This latter step depends critically on the availability of adequate energy sources which, we hold, will largely be driven by changes in the protoecosystem. Such changes may well have been in the form of punctuated shifts in resilience domain.

This is not, except for [4], a particularly new perspective, and is essentially similar to the 'bags of genes' model of Szathmary and colleagues (Szathmary and Demeter, 1987; Szathmary, 1989, 2006; Fontanari et al., 2005; Holmes, 2005) as well as Eigen's hypercycle compartment model (Eigen, 1996), but with hypercycles generalized to broad coevolutionary interactions. Our innovations lie in a highly formal use of the asymptotic limit theorems of information theory and in the systematic extension of phase transition techniques from statistical physics to information theory via 'biological' renormalizations.

Although the basic line of reasoning is fairly straightforward, a number of mathematical tasks need to be confronted. The first is to reconfigure the error catastrophe of Eigen's model in terms of average distortion. That done, the homology between information source uncertainty and free energy density can be extended to the rate distortion function, and, since the rate distortion limit is always a non-increasing convex function of the distortion, the basic model emerges as a kind of energy minimization near the error catastrophe. Pettini's (2007) topological hypothesis regarding the relation between topological shifts in structure and phase transition can then be applied, using characteristic 'biological renormalizations' to specify different forms of phase transition. A theory of 'all possible' Eigen models is direct. For analogs to the Gaussian channel, zero distortion – no error at all – is unattainable, requiring infinite energy. Such systems would, then, always be subject to some mutational variation, and, it follows, some type of natural selection.

The coevolutionary argument of the last chapter can then be applied to two or more vesicles interacting through a mutual information crosstalk produces the essential results: Again, the simplest possible system has two quasi-equilibrium points, one near the error limit, which is the low energy solution, and the other near zero error, the high energy solution.

The many different possible chemical strategies in this broad spread of possible solutions would themselves become higher order Darwinian individuals subject to the vagaries of evolutionary process in the context of large deviations, producing, on our planet, the familiar RNA-DNA chemistry basis of life. Other possibilities, however, seem particularly interesting, and may perhaps be observed *in vitro* or *ex planeta*.

5.2 Reconsidering the Eigen model

Following Campos and Fontanari (1999), the Eigen quasispecies model can be characterized in its binary version as follows:

A molecule is represented by a string of L digits, $(s_1, s_2, ..., s_L)$ with the variates s_α allowed to take only two different values, each of which represents a different type of monomer used to build the molecule. The concentrations x_i of molecules of type $i = 1, 2, ..., 2^L$ evolve in time according to the equations

$$dx_i/dt = \sum_j W_{i,j}x_j - \Phi(t)x_i$$

(5.1)

where $\Phi(t)$ is a dilution flux that keeps the total concentration constant, determined by the condition that $\sum_i dx_i/dt = 0$. Taking $\sum_i x_i = 1$ gives

$$\Phi = \sum_{i,j} W_{i,j}x_j.$$

(5.2)

The elements of $W_{i,j}$, the replication matrix, depend on the replication rate or fitness A_i of the strings of type i, as well as on the Hamming distance $d(i,j)$ between strings i and j. They are

$$W_{i,i} = A_i q^\nu$$

(5.3)

and

$$W_{i,j} = A_j q^{L-d(i,j)}[1-q]^{d(i,j)} i \neq j.$$

(5.4)

Here $0 \leq q \leq 1$ is the single-digit replication accuracy, assumed the same for all digits. Again, Leuthausser (1986) shows this model corresponds exactly to that of a two dimensional Ising system.

The famous figure on p.83 of Eigen (1996), taken from Swetina and Schuster (1982), shows a numerical realization of the model for $L = 50$. The fittest sequence has Hamming distortion $d = 0$, and is characterized as the 'master' sequence. As the error rate $(1 - q)$ increases from zero, the proportion of the population having the master sequence, x_0, declines, while those corresponding to other distortion values rise accordingly. The essential point, however, is that, for all values below the critical error threshold, the average across the quasispecies, all the x_d, what Eigen calls the 'consensus sequence', remains precisely the master sequence itself, even while the proportion constituting the master sequence itself falls. After the error threshold, however, that consensus average is lost, and the distributions become strictly random. The information of the master sequence has been dissipated.

Our interest is in reconsidering this effect from the perspective of the average distortion. As the error threshold is approached from below, the proportion corresponding the master sequence, having $d = 0$, declines, while the proportion of the population having $d > 0$ increases. Thus the average distortion, $D = \sum_j d_j x_j$, such that $\sum_i x_i = 1$, itself increases monotonically as the error rate approaches the error threshold, until a critical value of the average distortion D is reached.

The Rate Distortion Theorem, described in the Appendix, states that there is a minimum necessary rate of transmission of information, $R(D)$, through an information channel for any given average distortion D. That is, rates of transmission – channel capacities – greater than $R(D)$ are guaranteed to have average distortion less than D for any chosen distortion measure (Cover and Thomas, 1991; Dembo and Zeitouni, 1989).

We have translated the Eigen model from a focus on a single error threshold into one involving a critical value of average distortion in the transmission of information, and this will prove to be important for our subsequent modeling exercise.

We will, like others, attempt to solve Eigen's paradox by invoking a population of related mutants as constituting a parallel transmission channel having a sufficiently high capacity to ensure collective reproductive fidelity in spite of individual molecular reproductive errors. The 'consensus average' becomes the essential reproductive message, and different such populations separated into compartments or vesicles in the sense of Szathmary and Demeter (1987). These interact to become each other's principal environments, engaging in a crosstalk that enables the coevolutionary ratchet. Many different versions of this mechanism may have developed, becoming subject to evolutionary selection and chance extinction. Certainly both Eigen's compartment hypercycle (Eigen, 1996) and Szathmary's stochastic corrector (Szathmary and Demeter, 1987) solve the essential problem. Likely so too may a quite a plethora of different structures, or, as Szathmary (1989) comments, 'while it is true that hypercycles need compartments, do compartments need hypercycles?' Here we will attempt a kind of general compartment model, using a broad information-theoretic brush.

5.3 Capacity of a parallel channel

A trivial but important generalization of the simple channel without memory studied in Section 2.10 is the Parallel Channel. Consider a set of K discrete memoryless channels having capacities $C_1, ..., C_K$. Assume these are connected in parallel in the sense that for each unit of time an arbitrary symbol is transmitted and received over each channel. The input $\mathbf{x} = (x_1, ...x_K)$ to the parallel channel is a K-vector whose components are inputs to the individual channels, and the output $\mathbf{y} = (y_1, ..., y_K)$ is a K-vector whose components are the individual channel outputs. The capacity of the parallel channel is

$$C_{Total} = \sum_{i=1}^{K} C_i.$$

(5.5)

Thus the consensus sequence in a compartment or vesicle, which is clearly a parallel channel, can be transmitted at a much higher overall rate than is possible using individual replicators. This is no small matter.

5.4 Rate distortion dynamics

The rate distortion function of equation (8.8) in the Appendix can actually be calculated in many cases by using a Lagrange multiplier method – see Section 13.7 of Cover and Thomas (1991). For a simple Gaussian channel having zero mean noise with variance σ^2,

$$R(D) = 1/2 \log[\sigma^2/D], 0 \le D \le \sigma^2,$$

$$R(D) = 0, D > \sigma^2.$$

(5.6)

For this particular channel, zero distortion, no mutations at all, requires an infinite channel capacity, which, according to the homology with free energy density, requires infinite energy.

A second important observation is that *any* rate distortion function $R(D)$, following the arguments of Cover and Thomas, (1991, Lemma 13.4.1) is necessarily a *decreasing convex function* of D, that is, a reverse-J-shaped curve. This requirement, like the singularity of Gaussian-like channels at zero distortion, has profound consequences for replication dynamics.

It is possible to restate equations (2.7) and (2.8) in a manner which relates them closely to our central. First recall the classic relation between information source uncertainty and channel capacity:

$$H[\mathbf{X}] \leq C.$$

Next, the definition of channel capacity itself:

$$C \equiv \max_{P(X)} I(X|Y).$$

Finally, the definition of the rate distortion function, from equation (8.8):

$$R(D) = \min_{p(y,\hat{y});\sum_{(y,\hat{y})} p(y)p(y|\hat{y})d(y,\hat{y}) \leq D} I(Y,\hat{Y}).$$

$R(D)$ defines the minimum channel capacity necessary for average distortion D, placing a limits on information source uncertainty. Thus, we suggest distortion measures can drive information system dynamics.

We are led to propose, as a heuristic, that the dynamics of equations (2.7), (2.8), and (2.14) will, through the relation $H[\mathbf{X}] \leq C$, be constrained by the system as described in terms of a parametized rate distortion function. To do this, take the rate distortion function R as parametized, not only by the distortion D, but by some vector of variates $\mathbf{D} = (D_1, ..., D_k)$, for which the first component is the average distortion. The assumed dynamics are then driven by gradients in the rate distortion disorder defined as

$$S_R \equiv R(\mathbf{D}) - \sum_{i=1}^{k} D_i \partial R / \partial D_i,$$

(5.7)

leading to the deterministic and stochastic systems of equations

$$dD_j/dt = \sum_i L_{j,i} \partial S_R / \partial D_i$$

(5.8)

and

$$dD_t^j = L^j(t, D^1, ..., D^n)dt + \sum_i \sigma^{j,i}(t, D^1, ..., D^n)dB_t^i.$$

(5.9)

D^1 is the classic average distortion and the 'noise' terms dB_t^i are characterized by their quadratic variation. See the Mathematical Appendix for details.

A simple Gaussian channel, taking $\sigma^2 = 1$, has a Rate Distortion function

$$R(D) = 1/2 \log[1/D],$$

so that,

$$S_R(D) = R(D) - DdS_R/dD = 1/2 \log(1/D) + 1/2.$$

(5.10)

The simplest possible Onsager relation becomes

$$dD/dt \propto -dS_R/dD = \frac{1}{2D},$$

(5.11)

where $-dS_R/dD$ represents the force of the 'biochemical wind'. This has the solution

$$D \propto \sqrt{t}.$$

(5.12)

Similar results will accrue to any of the reverse-J-shaped relations which must inevitably characterize a rate distortion function. The implication is that simple RNA(-like) organisms – including RNA viruses – will inevitably be subject to a relentless biochemical evolutionary force, a powerful entropic wind, that can drive them very close to their critical mutation rates. These represent, in a fundamental sense, the minimum energy states possible to them as viable organisms. That is, in general, absent a contravening biological or other constraint requiring a constant energy influx,

$$D = f(t),$$

(5.13)

with $f(t)$ monotonic increasing in t.

It is not surprising, then, that so many RNA viruses are found close to their error thresholds, which do indeed constitute a powerful biological constraint (Holmes, 2003).

In the next section we will show how two or more interacting vesicles can, given enough energy, by virtue of a Red Queen dynamic, oppose this fierce chemical hurricane.

5.5 Rate distortion coevolution

We now extend equation (5.9) to an interacting system of vesicles, as in section 4.4, by allowing as parameters the inverse rate distortion functions representing interaction between vesicles. It is then possible to import wholesale the coevolutionary developments of sections 4.3-4.5, although the analog to figure 4.1 now involves simple reverse-J-shaped curves, since $R(D)$ is convex-decreasing in D.

A two-vesicle system then has two quasi-stable limit points, a low energy solution that must necessarily be near the error catastrophe, and a high energy state that may be near to, but never at, the zero error limit, depending on the available energy. Absent a relatively high energy source – predation, a new metabolic cycle, elementary photocapture, or the like – low error rate reproduction would be impossible, according to the model. This suggests some major ecological transformation in energy availability was a necessary condition for low error rate reproduction.

More complicated coevolutionary dynamics seem possible for larger systems, not only 'quasi-evolutionarily stable strategies', but even analogs to limit cycles or pseudorandom strange attractors, all having various rates of reproductive error.

A large deviations argument suggests that certain of these prebiotic chemical systems may, given time, undergo highly structured excursions among their various possible equilibrium or quasi-equilibrium modes, depending on available energy. The more viable of these collective modes having low error rates would then become subject to further evolutionary process, resulting in a condensation from RNA-like to DNA-like error correcting structures, where the various vesicles become ever more closely intertwined. Energy availability may indeed be a critical matter, with the probability of a 'large deviation' dependent on development or availability of new energy sources. The inverse perspective is that the sudden availability of a new energy source can drive large deviations, and hence evolutionary process. If one can argue that energy availability is enmeshed with large scale ecosystem resilience shifts, then this would be another example in which resilience drives evolution, albeit in prebiotic circumstances.

Our invocation of core concepts from information theory apparently marks a departure from current theorizing on these subjects. Applying the topological considerations of the earlier chapters, in particular the dynamical groupoid and directed homotopy arguments, a far richer theoretical structure is available, and will probably be required. One is, in spite of all the formal heavy lifting of this chapter, still somehow reminded of something like Sidney Harris' famous cartoon in which a white-coated scientist, standing in front of a blackboard filled with mathematical scribbles, turns to his audience, chalk in hand, saying "...and here a miracle occurs...".

Nonetheless, we have added to the store of theoretical tools useful in the study of prebiotic evolution and viral replication. The keys to further progress, as usual, however, seem to lie in more extensive and penetrating observational and laboratory studies.

Our simple model has implications beyond a prebiotic context. It is possible to formulate the contest between a highly mutating pathogen, operating at a low energy configuration, and a complex immune system, in terms of phenotype, rather than simply genotype, coevolution, in the sense of West-Eberhard (2003). Then the phenotype-phenotype 'two-vesicle' coevolutionary ratchet has, again, two quasi-fixed points, that is, fragmentation near a high variability phenotype, and immune-viral phenotype coalescence near a low variability phenotype. The latter will, ultimately, write an image of itself, in the sense of Adami et al. (2000), on the pathogen gene in spite of its high mutation rate, producing protected zones defining that phenotype. These may, in turn, be subject to targeting by a vaccine. The large deviations argument suggests, however, that phenotypic stability in pathogen-immune system interaction may itself undergo significant and degrading punctuated excursions under the influence of mesoscale resonance driven by embedding resilience domain shifts. The next chapter applies something of this perspective to the literal farming of human pathogens.

6

Farming human pathogens

Here we examine examples of the farming of coevolutionary systems, focusing on the mutually amplifying roles of large-scale psychosocial stress, economic structure, and reductionist interventions in the ecology and evolution of highly adaptive disease organisms. We find, in general, that population-level socioeconomic and other stressors, in synergism with reductionist interventions, are precisely suited to trigger mesoscale resonance coevolutionary resilience domain shifts affecting rapidly evolving pathogens. In an ideal world these changes would be the deliberate alteration of socioeconomic structure aimed at decreasing rates of infection and/or virulence, perhaps extending the utility of reductionist intervention. These case histories examine exactly contrary patterns.

6.1 Culture and the infection phenotype: a modeling exercise

Taking the perspectives of the earlier chapters, we can begin to model how population-directed, structured, psychosocial stress imposes an image of itself on the coevolutionary conflict between a highly adaptive chronic infection and the immune response.

As population-level structured stress appears a fundamental part of the biology of disease in human populations, this suggests the possibility that simplistic individual-oriented magic-bullet drug treatments, vaccines, and risk-reduction programs that do not address the fundamental living and working conditions which underlie disease ecology will fail to control many current epidemics. In addition, such reductionist interventions may go so far as to select for more holistic pathogens characterized by processes operating at multiple levels of biocultural organization.

R. Wallace et al., *Farming Human Pathogens*, DOI 10.1007/978-0-387-92213-3_6,
© Springer Science+Business Media, LLC 2009

6.1.1 Introduction

Earlier work in this direction (Wallace and Wallace, 2002; Wallace, 2002a) examined culturally-driven variation in HIV transmission and malaria pathology. HIV responds to immune challenge as an evolution machine, generating copious variation and hiding from counterattack in refugia at multiple scales of space, time, and population. *P. falciparum* engages in analogous rapid clonal antigenic variation, and cyto-adherence and sequestration in the deep vasculature, primary mechanisms for escaping from antibody-mediated responses of the host's immune system (Alred, 1998). Something much like the mutator phenotype (Thaler, 1999) or second order selection (Tenallion et al., 2001), by which the mechanisms mutations come about are themselves subjected to selection, appears to generate antigenic variation in the face of immune attack for a large class of pathogens. This could well be another version of the Baldwin effect.

Concomitantly, DiNoia and Neuberger (2002) outline the mechanisms by which the immune system's own antibody-producing B-cells engage in a second-order fine tuning of antibody production through somatic hypermutation, allowing organisms to respond quickly and effectively to pathogens that they have been exposed to previously (Gearhart, 2002).

Many chronic infections, particularly pathogens that cloak themselves in antigenic coats of many colors, are very often marked by distinct stages over the course of disease. HIV infections typically involve an initial viremia triggering an immune response that drives the virus into refugia during an extended asymptomatic period which, with the collapse of the immune system, ends in AIDS. Malaria's most evident stages are expressed as explosive outbursts of rapid parasite replication that facilitate insect-mediated transmission between hosts. HIV, malaria, and a third disease, tuberculosis, account for over five million deaths a year worldwide and exemplify the evolutionary success of multiple-stage chronicity as a life history strategy (Ewald, 2000; Villarreal, et al., 2000).

Here we analyze how pathogen life history stages represent a kind of co-evolutionary punctuation for chronic infection in the face of relentless immune and other selection pressures. For HIV that punctuation may arise from the direct interactions between the virus and the immune system response. In the case of malaria, it may result by means of a second order punctuation through the mutator mechanism (Thaler, 1999) associated with rapid antigenic variation. Elsewhere we have studied clonal selection in tumorigenesis from such a second order perspective (Wallace et al., 2003).

How can we characterize the interpenetration between antagonistic adaptive processes that defines disease dynamics? As described earlier, Adami et al. (2000) applied an information theoretic approach to conclude that genomic complexity resulting from evolutionary adaptation can be identified with the amount of information a gene sequence stores about its environment. Lewontin (2000) suggested something of a reverse process, in which environmental complexity represents the amount of information organisms introduce into their environment as a result of their collective actions and interactions. We propose modeling the interactions among information sources – generalized languages – provides a more faithful encapsulation of the interactive, multiscale nature of pathogen-immune dynamics than does the common differential equation predator-prey paradigm (e.g., Nowak and May, 2000).

Characterizing information sources as able to reflect their own context, as Adami et al. mapped out, we have applied a rate distortion argument in the context of imposed renormalization symmetry to obtain evolutionary punctuated equilibrium, and can use the more general Joint Asymptotic Equipartition Theorem (JAEPT) to conclude that pathogenic adaptive response and coupled cognitive immune challenge will be jointly linked in chronic infection, and subject to a transient punctuated interpenetration very similar to evolutionary punctuation. Multiple punctuated transitions, perhaps of mixed order, may well constitute shifts to the different stages of chronic infection.

Examining paths in parameter space for the renormalization properties of such transitions – the universality class tuning of chapter 3 – produces a second order punctuation in the rate at which the selection pressure of the immune system imposes a distorted image of itself onto pathogen structure. This is our version of the mutator or Tenallion et al.'s second order selection.

Recognizably similar matters have long been under scrutiny: interactions between the central nervous system (CNS) and the immune system, and between genetic heritage and the immune system have become academically codified through journals with titles such as *Neuroimmunology* and *Immunogenetics*. Elsewhere (Wallace and Wallace, 2002) we introduced another complication by arguing that the culture in which humans are socially embedded also interacts with individual immune systems to form a composite entity that we labeled an *immunocultural condensation* (ICC). It is, we will argue here, the joint entity of immune, CNS, and embedding sociocultural cognition that engages in orders of punctuated interpenetration with an adaptive chronic infectious challenge. Similar arguments are already in the French literature (e.g., Combes, 2000).

Included among the most damaging cultural inputs on immune system function are the long-term psychosocial stresses of war, oppression, and discrimination imposed by one population on another. If valid, the paradigm has fundamental consequences for concepts of human biology. While Diamond (1997) and others (Crosby, 1986; Hughes,2001) popularized ecological explanations of human history, the paradigm presented here suggests investigation in something of the other direction, at the means by which human history shapes biological ontogeny, often through punctuated processes of mesoscale resonance.

The paradigm would appear to have practical implications as well. Interpenetrations among pathogens, the immune system's response, and the embedding culture in which individuals find themselves would greatly color the success of the kinds of individual-level disease interventions largely pursued today. Reductionist interventions – drug regimens, vaccines, risk reduction programs – aimed at holistic diseases, defined by myriad processes operating at multiple scales of time and space both within and without individuals, are likely to fail. Furthermore, what successes reductionist interventions have had against reductionist diseases may very well select for holistic diseases able to dilute or deflect the effectiveness of interventions pursued at single scales alone.

There are some general considerations. First, the information theory approach we have adopted in this book is notorious for providing existence theorems whose representation, to use physics jargon, is arduous. For example, although the Shannon Coding Theorem implied the possibility of highly efficient coding schemes as early as 1949, it took more than forty years for practical turbo codes to be created. The program we propose is unlikely to be any less difficult.

Second, we are invoking information theory variants of the fundamental limit theorems of probability. These are independent of exact mechanisms, but, as necessary conditions, constrain the behavior of those mechanisms. For example, although not all processes involve long sums of independent stochastic variables, those that do, regardless of the individual variable distribution, collectively follow a Normal distribution as a consequence of the Central Limit Theorem. Similarly, the games of chance in a Las Vegas casino are all quite different, but nonetheless the success of strategies for playing them is strongly and systematically constrained by the Martingale Theorem, regardless of game details. Languages-on-networks and languages-that-interact, as a consequence of the limit theorems of information theory, will be subject to necessary-condition regularities of punctuation and generalized Onsager relations, regardless of detailed mechanisms, as important as the latter may be.

Finally, just as parametric statistics are imposed, at least as a first approximation, on sometimes questionable experimental situations, relying on the robustness of the Central Limit Theorem to carry us through, we will invoke here a similar heuristic approach for the information theory limit theorems we define.

We begin with a reiteration and reinterpretation of some results from chapter 3.

6.1.2 Universality class tuning

Here we again iterate the general argument of chapter 3 onto the process of phase transition itself, obtaining Tenallion's second order selection – the mutator – in a natural manner.

We suppose that a structured environment, which we take itself to be an appropriately regular information source Y – e.g., the immune system, or more generally, for humans the immunocultural condensation (ICC) – engages a modifiable system – e.g., a pathogen – through selection pressure.

The ICC begins to write itself on the pathogen's genetic sequences or protein residues in a distorted manner permitting definition of a mutual information $I[K]$ splitting criterion according to the Rate Distortion or Joint Asymptotic Equipartition Theorems. K is an inverse coupling parameter between system and environment. According to our development, at punctuation – near some critical point K_C – the systems begin to interact very strongly indeed, and we may write, near K_C, taking as the starting point the simple physical model of section 3.4,

$$I[K] \approx I_0[\frac{K_C - K}{K_C}]^\alpha.$$

For a physical system α is fixed, determined by the underlying universality class. Here we will allow α to vary, and to itself respond explicitly to selection pressure.

Normalizing K_C and I_0 to 1, we obtain,

$$I[K] \approx (1 - K)^\alpha.$$

(6.1)

To repeat, the horizontal line $I[K] = 1$ corresponds to $\alpha = 0$, while $\alpha = 1$ gives a declining straight line with unit slope which passes through 0 at $K = 1$. Consideration shows there are progressively sharper transitions between the necessary zero value at $K = 1$ and the values defined by this relation for $0 < K, \alpha < 1$. The rapidly rising slope of transition with declining α is, we assert, of considerable significance.

Again, the instability associated with the splitting criterion $I[K]$ is defined by

$$Q[K] \equiv -KdI[K]/dK = \alpha K(1 - K)^{\alpha-1},$$

(6.2)

and is singular at $K = K_C = 1$ for $0 < \alpha < 1$. And again we interpret this to mean that values of $0 < \alpha \ll 1$ are highly unlikely for real systems, since $Q[K]$, in this model, represents a kind of energy barrier for information systems.

On the other hand, smaller values of α mean that the system is far more efficient at responding to the adaptive demands imposed by the embedding structured ecosystem, since the mutual information which tracks the matching of internal response to external demands, $I[K]$, rises more and more quickly toward the maximum for smaller and smaller α as the inverse coupling parameter K declines below $K_C = 1$. That is, *systems able to attain smaller α are more adaptive than those characterized by larger values*, in this model, but smaller values will be hard to reach, and can probably be done so only at some considerable physiological or other cost, an energy argument similar to that of the previous chapter.

The more biologically realistic renormalization strategies given in chapter 3 produce sets of several parameters defining the universality class, whose tuning gives behavior much like that of α in this simple example.

We can formally iterate the phase transition argument on this calculation to obtain our version of the mutator, focusing on paths of universality classes.

6.1.3 The adaptive mutator

Suppose the renormalization properties of a biological or social language-on-a network system at some 'time' k are characterized by a set of parameters $A_k \equiv \alpha_1^k, ..., \alpha_m^k$. Fixed parameter values define a particular universality class for the renormalization. We suppose that, over a sequence of 'times', the universality class properties can be characterized by a path $x_n = A_0, A_1, ..., A_{n-1}$ having significant serial correlations which, in fact, permit definition of an adiabatically piecewise memoryless ergodic information source associated with the paths x_n. We call that source $\mathbf{X}$.

We further suppose, as described earlier, that external selection pressure is also highly structured – e.g., the cognitive immune system or, in humans, the ICC – and forms another information source $\mathbf{Y}$ which interacts not only with the system of interest globally, but specifically with its universality class properties as characterized by $\mathbf{X}$. $\mathbf{Y}$ is necessarily associated with a set of paths y_n.

We pair the two sets of paths into a joint path, $z_n \equiv (x_n, y_y)$ and invoke an inverse coupling parameter, K, between the information sources and their paths. This leads, by the arguments above, to phase transition punctuation of $I[K]$, the mutual information between $\mathbf{X}$ and $\mathbf{Y}$, under either the Joint Asymptotic Equipartition Theorem or under limitation by a distortion measure, through the Rate Distortion Theorem (Cover and Thomas, 1991). The essential point is that $I[K]$ is a splitting criterion under these theorems, and thus partakes of the homology with free energy density which we have invoked above.

Activation of universality class tuning, our version of the mutator, then becomes itself a punctuated event in response to increasing linkage between organism – the pathogen – and externally imposed selection or other pressure – responses of the ICC. Mutation rates become a function of the relationship between the ICC and the pathogen, above and beyond environmental insult alone.

Thaler (1999) has suggested that the mutagenic effects associated with a cell sensing its environment and history could be as exquisitely regulated as transcription. Our invocation of the Rate Distortion or Joint Asymptotic Equipartition Theorems in address of the mutator necessarily means that variation comes to significantly reflect the grammar, syntax, and higher order structures of the embedding processes. This involves far more than a simple colored noise – stochastic excursions about a deterministic spine – and most certainly implies the need for exquisite regulation. Our information theory argument here converges with Thaler's speculation.

In the same paper Thaler further argues that the immune system provides an example of a biological system which ignores conceptual boundaries that separate development from evolution. While evolutionary phenomena are not cognitive in the sense of the immune system (Cohen, 2000), they may still partake of a significant interaction with development, in which the very reproductive mechanisms of a cell, organism, or organization become closely coupled with structured external selection pressure in a manner recognizably analogous to 'ordinary' punctuated evolution.

That is, we argue the staged nature of chronic infectious diseases like HIV and malaria represents a punctuated version of biological interpenetration, in the sense of Lewontin (2000), between a cognitive 'immunocultural condensation' and a highly adaptive pathogen. We further suggest that this punctuated interpenetration may have both first (i.e., direct) and second order characteristics, involving cross interactions between direct cognitive effects of the immune system or immunocultural condensation, or, more generally, of the ICC and the mutator mechanisms of both the immune system and its pathogen targets.

Another path to the mutator might be through a second order iteration similar to that just above, but focused on the parameters defining the universality class distributions of section 3.3.

6.1.4 Population stress and pathogen response

As we discuss elsewhere (Wallace and Wallace, 2002; Wallace, 2002a), structured psychosocial stress directed at populations, by policy choice or as unforeseen consequence, constitutes a determining context for immune cognition or, more generally, the immunocultural condensation. We wish to analyze the way structured stress affects the interaction between the cognitive ICC and an adaptive mutator, the principal line of defense against the ICC for a large class of highly successful pathogens. To do this we must extend our theory to three interacting information sources, briefly reiterating the argument of Section 2.4.

The Rate Distortion and Joint Asymptotic Equipartition Theorems are generalizations of the Shannon-McMillan Theorem which examine the interaction of two information sources, with and without the constraint of a fixed average distortion. We conduct one more iteration, and require a generalization of the SMT in terms of the splitting criterion for triplets as opposed to single or double stranded patterns. The tool for this is at the core of what is termed network information theory (Cover and Thomas, 1991, Theorem 14.2.3), leading to equations (2.1) and (2.2). We briefly review that development.

Suppose we have three (adiabatically piecewise stationary) ergodic information sources, Y_1, Y_2 and Y_3. We assume Y_3 constitutes a critical embedding context for Y_1 and Y_2 so that, given three sequences of length n, the probability of a particular triplet of sequences is determined by *conditional probabilities with respect to Y_3*:

$$P(Y_1 = y_1, Y_2 = y_2, Y_3 = y_3) =$$

$$\Pi_{i=1}^n p(y_{1i}|y_{3i})p(y_{2i}|y_{3i})p(y_{3i}).$$

(6.3)

That is, Y_1 and Y_2 are, in some measure, driven by their interaction with Y_3

Then, as with previous analyses, triplets of sequences can be divided by a splitting criterion into two sets, having high and low probabilities respectively. For large n the number of triplet sequences in the high probability set will be determined by the relation (Cover and Thomas, 1992, p. 387),

$$N(n) \propto \exp[nI(Y_1; Y_2|Y_3)],$$

(6.4)

where splitting criterion is given by equation (2.1),

$$I(Y_1; Y_2|Y_3) \equiv$$

$$H(Y_3) + H(Y_1|Y_3) + H(Y_2|Y_3) - H(Y_1, Y_2, Y_3).$$

We can then examine mixed cognitive/adaptive phase transitions analogous to learning plateaus (Wallace, 2002b) in the splitting criterion $I(Y_1, Y_2|Y_3)$, which characterizes the synergistic interaction between structured psychosocial stress, the ICC, and the pathogen's adaptive mutator. These transitions delineate the various stages of the chronic infection, which are embodied in the slowly varying APSE phase between transitions. Again, our results are closely analogous to the Eldredge-Gould treatment of evolutionary punctuated equilibrium in evolution.

We can, if necessary, extend this model to any number of interacting information sources, $Y_1, Y_2, ..., Y_s$ conditional on an external context Z in terms of a splitting criterion defined by equation (2.2):

$$I(Y_1; ...; Y_s|Z) = H(Z) + \sum_{j=1}^{s} H(Y_j|Z) - H(Y_1, ..., Y_s, Z),$$

where the conditional Shannon uncertainties $H(Y_j|Z)$ are determined by the appropriate direct and conditional probabilities.

6.1.5 The phenotype coevolution ratchet

We have so far focused this treatment on complex parasites such as malaria which may have mutator mechanisms determining behavior of their antigentic coat of many colors. A simplified analysis can also be applied directly to HIV, which, as a kind of evolution machine, seems to engage in endless, rapid, direct mutation, and, at broader temporal scales, recombination. The essential argument regarding RNA viruses is that the high error rate inherent to viral replication is very nearly the lowest possible energy state consonant with quasi-species survival, according to the rate distortion argument of chapter 5. Nonetheless, a virus does not exist alone. It functions in cooperation with, as well as in conflict with, the host organism. Some RNA viruses, for example poliomylitus and measles, have high error rate replication, but a (nearly) fixed phenotype, permitting effective vaccination.

The development of Chapter 5 suggests that the coevolutionary ratchet between virus and immune system can be generalized to phenotype-phenotype interactions. That is, to the competition between the virus and the cognitive immune system. At the level of the pathogen-host system, the possible coevolutionary stable states are either a highly variable phenotype-phenotype conflict near a critical point, or else a fixed, phenotype-phenotype quasi-equilibrium point at the low variability end of the ratchet.

This dichotomy appears to be extended to structurally more complex pathogens, for example, malaria, which, although it is not near some evolutionary error catastrophe, has apparently nonetheless been ratcheted down to the high variability equilibrium point for phenotype coevolution.

The cognitive nature of the immune system may play an important role here. SIV, the simian version of HIV, in long-evolved relations with a host

species, exists at high blood titre without eliciting an inflammatory response
(Gordon et al., 2005). In terms of the cognitive model of section (1.2), SIV
antigen has been relegated to the 'B_0' model of 'not recognized' rather than
the 'B_1' mode of immune attack. One speculates as to the possible importance of cognitive gene expression in determining more complex phenotypic
coevolutionary processes. An example is perhaps found in the recent work of
Ley et al. (2008) on the evolution of mammals and their gut microbes. They
conclude that the tolerance of the immune system to gut microbes is a basal
trait in mammal evolution.

As stated at the end of chapter 5, however, a large deviations argument
suggests that external mesoscale ecosystem shifts might well drive a relatively
stable host-pathogen relationship from low to high variability, not a good
thing.

6.1.6 Implications of the model

Scientific enterprise encompasses the interaction of facts, tools, and theories,
all embedded in a path-dependent political economy that seems as natural to
us as air to a bird. Molecular biology, Central Limit Theorem statistics, and
19th century mathematics, presently provide the reductionist tool kit most
popular in the study of immune function and disease process. Many essential
matters related to the encompassing social, economic, and cultural matrix so
fundamental to human biology are simply blindsided, and one is reminded,
not very originally, of the joke about the drunk looking for his missing car
keys under a street lamp, "because the light here is better."

The asymptotic limit theorems of probability beyond the Central Limit
Theorem, in concert with related formalism adapted from statistical physics,
would seem to provide new tools. We think these can generate theoretical
speculations of value in obtaining and interpreting empirical results about
infection and immune process.

Our model explicitly invokes the possibility of synergistic interaction between the selection pressure of the immunocultural condensation (ICC) that
characterizes human immune response and the variable antigenic coat of an
established pathogen population, particularly in the context of embedding
patterns of structured psychosocial stress which, to take a Rate Distortion
perspective, can literally write an image of itself onto that interaction. The
ICC, through immune hypermutation and the choice of immune response pursued, may engage in its own second order selection. What results are first,
second, and possibly mixed, order interpenetrations, in which the ICC and
pathogens constitute each other's selection pressure and selected structure,
an interaction that may become a distorted image of enfolding patterns of
socioeconomically, historically, and politically determined psychosocial stress.
As the evolutionary anthropologist Robert Boyd put succinctly, Culture is as
much a part of human biology as the enamel on our teeth (Boyd and Richerson

1995; Richerson and Boyd, 2004). It follows that any efforts to characterize and respond to threats to human biology need account for culture's roles.

Human chronic infection cannot, in particular, be simply abstracted as a matter of conflict between the pathogen and the immune system alone. Indeed, the concept of an immune system 'alone' has no meaning within our model, in stark contrast with, for example, the well-stirred Erlenmeyer flask predator-prey population dynamics of Nowak and May (2000). The cells of the immune system comprise only the point of a long biocultural sword aimed at the throats of most infections.

Individual and collective history, socioeconomic structure, psychosocial stress and the resulting emotional states, may not be mere adjuncts to what is termed basic science in the medical journals. Rather, they may be as much a part of basic human biology as T-cells. Magic bullet vaccines, therapeutic drugs, or highly-focused medicalized social interventions against HIV disease and other mutagenic parasites – approaches that inherently cannot reckon with socioeconomic, historical, and cultural determinants of health and illness – will likely largely fail as they are overwhelmed by a combination of relentless pathogen adaptation, cross-population variation in immune cognition, and a globalized travel network that increasingly confronts host populations with myriad pathogen variants. For chronic infections like HIV and malaria, individual level or limited social network intervention strategies which neglect larger embedding context, and the history of that context, embody a grossly unreal paradigm of basic human biology.

We know that some social systems have succeeded in controlling malaria through, for example, persistent and highly organized programs of insect vector control. For HIV, humans are both vector and host. The larger social context, then, plays a fundamental role in the individual- and population-level decisions that promote or decelerate the HIV epidemic (D. Wallace and R. Wallace 1998; Schoepf et al. 2000). The biological consequences of ignoring the larger context are devastating, above and beyond the awful human cost of the epidemic.

R.G. Wallace (2004) suggests that, alone, individual-level antiretroviral treatment of the HIV epidemic may constitute a selection pressure forcing evolutionary changes in HIV life history, including, in one albeit remote possibility, a more rapid onset of AIDS. A key result, however, is that increasing infection survivorship and decreasing the transition rate from the asymptomatic stage to AIDS, as drug regimens aim to do, may induce the greatest increase in infection population growth. Because infection survivorship is physiologically enmeshed with host survivorship the asymptomatic stage becomes under the drug regimens a demographic shield against epidemiological intervention. In other words, HIV may use processes at one level of biocultural organization to defend itself against cures directed at it at other levels. Any successful intervention, then, must display a comparable multidimensionality.

Cartesian reductionism internalizes causality by assuming the whole of any phenomenon is a sum of its parts. Despite its successes, many (Wimsatt, 1980;

Bignami, 1982; Levins and Lewontin, 1985; Mayr, 1996; Levins, 1998; Oyama et al., 2001; Van Regenmortel and Hull, 2002; Gould 2003) have pointed out the problems with the reductionist approach in the study of biological phenomena, including of disease. Reductionism's widespread application, even on problems that do not yield to its approaches, is in part an outgrowth of social decisions about the role and nature of science.

Our work, here and cited, suggests a further complication. The consequences of reductionism's failures do not merely include mischaracterizing epidemics. The nature of study itself can affect the evolution of pathogens. The reductionist approach may very well, through a mesoscale resonance-driven microbial ecosystem resilience domain shift, *select for* holistic or dialectical responses on the part of pathogens. Reductionism's wholesale application, while succeeding against diseases such as polio and smallpox, welcome developments notwithstanding, may select for diseases that are characterized by complex sociogeographies, multiple hosts, and multidimensional interactions across scale. The HIV, malaria, and tuberculosis epidemics, as we have discussed, are obvious examples of holistic pathogens. In industrial countries, heart disease, cancer, and obesity take their toll; so-called diseases of affluence the poorest and most marginalized typically suffer the worst (Wallace et al., 2003; Wallace, 2005a). The ecology literature tells us sources of mortality compete. While pharmaceuticals, surgery, and individual-level risk reduction interventions control reductionist threats – additive sources of mortality both within individuals and populations – the pathogenic playing field appears now tilted towards holistic diseases we are largely unable to address because of the restricted scientific and medical practices pursued.

Our model raises the possibility of effective integrated pathogen management (IPM) programs through synergistic combinations of social, ecological, and medical interventions. IPM far transcends 'medical' strategies that amount to little more than a kind of pesticide application, an approach increasingly abandoned in agriculture as simply inadequate to address pathogen evolutionary strategies.

Prospects for studying immunocultural condensation and implementing a related IPM appear both exciting and difficult. New modes of intervention need involve new means of modeling basic biology. While we can model the interaction of first and second order phenomena in the context of structured stress using network information theory, it is difficult to envision interaction between second order 'tuning' processes, or the mechanics of even higher order effects: can we continue to 'tune the tuners' in a kind of idiotypic hall of mirrors? The mathematics would be straightforward, but the corresponding molecular biology would have to be subtle indeed. Higher order interpenetration – mutating the mutator – may be observable in certain isolated circumstances, for example in the interplay between B-cell somatic hypermutation and a pathogen's hypervariable membrane proteins. More likely some version of rate distortion manifold retina-like focus operates. Clearly much work is needed to trace the connections among the culture-specific and cognitive nature of the immune system, pathogen adaptation, the information dynamics of their interaction, the molecular manifestations of those dynamics, and the particularities of intervention.

As a first effort toward testing the proposed relations among the ICC and disease, we next qualitatively apply our paradigm to characterizing specific pathogens and the socioecological contexts in which they evolve.

6.2 Culture and the infection phenotype: case histories

6.2.1 Introduction

Following closely Wallace and Wallace (2002), we begin with a reconsideration of some implications of the Atlan-Cohen perspective on immune cognition (Cohen, 1992, 2000; Atlan and Cohen, 1998) for understanding the role of culture in the phenotypic expression of infectious disease, and the implications for vaccine strategies when simple elicitation of sterilizing immunity fails. This will provide an introduction to more complicated circumstances in which culture, and the policies which derive organically from it, can actually drive pathogen evolution.

The Atlan-Cohen view takes on special importance in the context of recent work by Nisbett et al. (2001) showing clearly that cognition in the central nervous system (CNS) is not universal, but rather differs fundamentally for populations with different cultural systems. We propose the immune system too may be a culture-specific condensation of sociocultural and biological cognition, in the same sense that neuroimmunology and immunogenetics describe the condensation of CNS and genetic 'languages' with immune function. Modifying Boyd's aphorism about culture described previously, we propose that culture is as much a part of the human immune system as T-cells. It follows that successful vaccine strategies where the smallpox model fails most likely must take such immunocultural condensation into account. In this introductory section we reinterpret recent studies of West African cultural variation in immune response to malaria, and in the efficacy of interventions against it. We also review similar US cultural variation in HIV transmission. The approach neither reifies 'race' nor, as in much of the biomedical literature, denies the burdens of social and political histories.

Malaria and HIV are major causes of morbidity and mortality for which no vaccine strategy has produced sterilizing immunity. Malaria has a complicated parasite life cycle with multiple and often changing antigens, and HIV is an evolution machine. Indeed, many, if not most, infectious diseases and malignancies have basic ecological and life-cycle factors that obviate simple effective vaccination on the smallpox model.

Such complications are increasingly under scrutiny. For example interactions between the central nervous system (CNS) and the immune system, and between the genetic heritage and the immune system have become officially recognized and academically codified through journals with titles such as *Neuroimmunology* and *Immunogenetics*. Another complication, described in the previous section, recognizes that the culture in which humans are socially embedded also interacts with individual immune systems to form a composite entity that might well be labeled an immunocultural condensation, (ICC).

In the light of the ICC we reinterpret recent observations of culturally-specific immune response to malaria in West Africa, and to heterosexual AIDS in the US.

6.2.2 Genes, cognition, and culture

Increasingly, biologists excoriate simple genetic reductionism which neglects the role of environment. Lewontin (2000), for example, explains that genomes are not 'blueprints,' as genes do not 'encode' for phenotypes. Organisms are instead outgrowths of fluid, conditional interactions between genes and their environments, as well as developmental 'noise.' Organisms, in turn, shape their environments, generating what Lewontin terms a triple helix of cause and effect. Such interpenetration of causal factors may be embodied by an array of organismal phenomena, including, as we shall discuss, a fourth branch to the Lewontin helix, that is, culture's relationships with the brain, the immune system, and the ecology of infectious disease. We propose reinterpreting immune function in this light, with profound implications for medical and public health interventions for infectious diseases where the smallpox model fails.

The current vision of human biology among evolutionary anthropologists is consistent with Lewontin's analysis and is summarized by Durham (1991). Durham argues that genes and culture are two distinct but interacting systems of inheritance within human populations. Information of both kinds has influence, actual or potential, over behaviors which creates a real and unambiguous symmetry between genes and phenotypes on the one hand, and culture and phenotypes on the other. Genes and culture are best represented as two parallel lines or 'tracks' of hereditary influence on phenotypes.

A goodly part of hominid evolution can be characterized as an interweaving of genetic and cultural systems. Genes came to encode increasing hypersociality, learning, and language skills. The most successful populations displayed increasingly complex structures that better aided in buffering the local environment (Bonner, 1980). Every successful human population seems to have a core of tool usage, sophisticated language, oral tradition, mythology and music, focused on relatively small family/extended family groupings of various forms. More complex social structures are build on the periphery of this basic genetic-cultural object (Richerson and Boyd, 2004).

At the level of the individual, the genetic-cultural object appears to be mediated by what evolutionary psychologists postulate are cognitive modules within the human mind (Barkow et al., 1992). At the risk of reifying a preformationist ontology, each module was shaped by natural selection in response to specific environmental and social conundrums Pleistocene hunter-gatherers faced. One set of such domain-specific cognitive adaptations addresses problems of social interchange (Cosmides and Tooby, 1992). Regardless of the exact origins of the human mind, the human species' very identity may rest, in part, on its unique evolved capacities for social mediation and cultural transmission.

The brain-and-culture condensation has been adopted as a kind of new orthodoxy in recent studies of human cognition. For example Nisbett et al. (2001) review an extensive literature on empirical studies of basic cognitive differences between individuals raised in what they call 'East Asian' and 'Western' cultural heritages. They view Western-based pattern cognition as 'analytic' and East-Asian as 'holistic.' Nisbett et al. (2001) find that

[1]. Social organization directs attention to some aspects of the perceptual field at the expense of others.

[2]. What is attended to influences metaphysics.

[3]. Metaphysics guides tacit epistemology, that is, beliefs about the nature of the world and causality.

[4]. Epistemology dictates the development and application of some cognitive processes at the expense of others.

[5]. Social organization can directly affect the plausibility of metaphysical assumptions, such as whether causality should be regarded as residing in the field vs. in the object.

[6]. Social organization and social practices can directly influence the development and use of cognitive processes such as dialectical vs. logical ones.

Nisbett et al. (2001) conclude that tools of thought embody a culture's intellectual history, that tools have theories build into them, and that users accept these theories, albeit unknowingly, when they use these tools.

We argue that the condensation between culture and both the gene and the brain described here may also be found for the immune system. Next we briefly review some implications of the Atlan-Cohen arguments regarding immune cognition.

6.2.3 Immune cognition and culture

Section 1.2.1 examined the Atlan/Cohen view of immune cognition at some length. As we have shown earlier, it is possible to give Atlan and Cohen's language metaphor of meaning-from-response a precise information-theoretic characterization, and to place that characterization within a context of recent developments which propose the coevolutionary mutual entrainment of different information sources to create larger metalanguages with the originals as subdialects. This work, a formalism based on the Large Deviations Program of applied probability, permits treating gene-culture and brain-culture condensations using a unified conceptual framework of information source coevolutionary condensation. Cohen's immune cognition model suggests, then, the possibility that human culture and the human immune system may be jointly convoluted. That is, there would appear to be, in the sense of the gene-culture and brain-culture condensations of the previous section, an immune-culture condensation as well. To neuroimmunology and immunogenetics we add 'immunocultural condensation'.

The evolutionary anthropologists' vision of the world implies language, culture, gene pool, and individual CNS and immune cognition are intrinsically melded and synergistic. We propose that where the smallpox vaccine model fails, culture and immune cognition may become a joint entity, determining, in considerable measure, the kind of vaccine strategy which may be effective. This effect may be confounded – and even masked – by the distinct population genetics often associated with linguistic and cultural isolation.

Africa contains great cultural and genetic diversity, suggesting the need for severe local refining and monitoring of any vaccine strategy. Traditional 'case-control' studies can, in fact, be profoundly compromised by linguistic and cultural differences which are convoluted with an associated genetic divergence that may be a simple marker of such difference rather than its cause. Similarly, the US, as a nation of immigrants, encompasses considerable cultural and genetic diversity, even in the context of both de-jure and de-facto deculturation.

In sum, population differences of immune function heretofore attributed to genetic factors alone may, rather, represent differences in immune cognition driven by, or, through the proposed ICC, synergistic with, profound cultural differences.

We reinterpret recent observations on malaria in Burkina Faso and heterosexual AIDS in New Jersey from this perspective.

6.2.4 Malaria and the Fulani

Modiano et al. (1996, 1998, 2001) have conducted comparative surveys on three roughly co-resident West African ethnic groups – which they describe as 'sympatric' – exposed to the same strains of malaria. The Fulani, Mossi, and Rimaibe live in the same conditions of hyperendemic transmission in a Sudan savanna area northeast of Ouagadougou, Burkina Faso. The Mossi and Rimaibe are Sudanese Negroid populations with a long tradition of sedentary farming, while the Fulani are nomadic pastoralists, partly settled and characterized by non-Negroid features of possible Caucasoid origin.

Parasitological, clinical, and immunological investigations showed consistent interethnic differences in *P. falciparum* infection rates, malaria morbidity, and prevalence and levels of antibodies to various *P. falciparum* antigens. The data point to a remarkably similar response to malaria in the Mossi and Rimaibe, while the Fulani are clearly less parasitized, less affected by the disease, and more responsive to all antigens tested. No difference in the use of malaria protective measures was demonstrated that could account for these findings. Known genetic factors of resistance to malaria showed markedly *lower* frequencies in the Fulani (Modiano et al., 2001). The differences in the immune response were not explained by the entomological observations, which indicated substantially uniform exposure to infective bites.

Modiano et al. (1996) conclude that sociocultural factors do not seem to be involved, and that the available data support the existence of unknown genetic factors, possibly related to humoral immune responses, determining interethnic differences in the susceptibility to malaria.

In spite of later finding the Fulani in their study region have significantly reduced frequencies of the classic malaria-resistance genes compared to the other 'sympatric' ethnic groups, Modiano et al. (2001) again conclude that their evidence supports the existence in the Fulani of unknown genetic factor(s) of resistance to malaria.

This vision of the world carries consequences, seriously constraining interpretation of the efficacy of interventions. Modiano et al. (1998) conducted an experiment in their Burkina Faso study zone involving the distribution of permethrin-impregnated curtains (PIC) to the three populations, with markedly different results:

> " The PIC were distributed in June 1996 and their impact on malaria infection was evaluated in [the three] groups whose baseline levels of immunity to malaria differed because of their age and ethnic group. Age- and ethnic-dependent efficacy of the PIC was observed. Among Mossi and Rimaibe, the impact (parasite rate reduction after PIC installation with respect to the pre-intervention surveys) was 18.8 % and 18.5 %, respectively. A more than two-fold general impact (42.8 %) was recorded in the Fulani. The impact of the intervention on infection rates appears positively correlated with the levels of anti-malaria immunity..."

Most critically, Modiano et al. (1998) conclude from this experiment that the expected complementary role of a hypothetical vaccine is stressed by these results, which also emphasize the importance of the genetic background of the population in the evaluation and application of malaria control strategies.

While we fully agree with the importance of the results for a hypothetical vaccine, much in the spirit of Lewontin (2000) we beg to differ with the ad hoc presumptions of genetic causality, which paper over alternatives involving environment and development consistent with these observations.

The medical anthropologist Andrew Gordon has published a remarkable study of Fulani cultural identity and illness (Gordon, 2000):

> "Cultural identity – who the Fulani think they are – informs thinking on illnesses they suffer. Conversely, illness, so very prevalent in sub-Saharan Africa, provides Fulani with a consistent reminder of their distinctive condition... How they approach being ill also tells Fulani about themselves. The manner in which Fulani think they are sick expresses their sense of difference from other ethnic groups. Schemas of [individual] illness and of collective identity draw deeply from the same well and web of thoughts... As individuals disclose or conceal illness, as they discuss illness and the problem of others, they reflect standards

of Fulani life – being strong of character not necessarily of body, being disciplined, rigorously Moslem, and leaders among lessors... to be in step with others and with cultural norms is to have pride in the self and the foundations of Fulani life."

The Fulani carried the Islamic invasion of Africa into the sub-Sahara, enslaving and deculturing a number of ethnic groups, and replacing the native languages with their own. This is much the way African Americans were enslaved, decultured, and taught English.

As Gordon puts it,

" 'True Fulani' see themselves as distinguished by their aristocratic descent, religious commitments, and personal qualities that clearly differ from lowland cultivators. Those in the lowland are, historically, Fulani subjects who came to act like and speak Fulani, but they are thought to be without the right genealogical descent. The separation between pastoralists and agriculturists repeats itself in settlements across Africa. The terms vary from place to place in Guinea, the terms are Fulbhe for the nobles and the agriculturalist Bhalebhe or Maatyubhee; in Burkina Faso, Fulbhe and the agricultural Rimaybhe; and in Nigeria, the Red Fulani and the agricultural Black Fulani... The schemas for the Fulani body describe the differences between them and others. These are differences that justify pride in being Fulani and not Bhalebhe, Maatyubhe, Rimaybhe, or Black Fulani. In Guinea, the word 'Bhalebhe' means 'the black one'. The term 'Bhalebhe' carries the same meaning as 'Negro' did for Africans brought to North America. It effaces any tribal identity...

The control a Fulani exercises over the body is an essential feature of 'the Fulani way.' Being out of control is shameful and not at all Fulani-like... To act without restraint is to be what is traditionally thought of as Bhalebhe...

Being afflicted with malaria – and handling it well – is a significant proof of ethnicity. How Fulani handle malaria may be telling. What they lack in physical resistance to disease they make up in persistence. Though sickly, Fulani men only reluctantly give into malaria and forgo work. To give into physical discomfort is not *dimo*. When malaria is severe for a man he is likely not to succumb to bed, but instead to sit outside of his home socializing."

Parenthetically, many primate studies (e.g., Schapiro et al., 1998) show that dominance rank, an important psychosocial factor, strongly and positively affects immune response in a stable social setting, while a vast body of parasitological observation and theory (e.g., Crofton, 1971) shows the 'overdispersion' of parasites within affected populations – i.e., relative concentration – is closely but inversely related to social dominance.

Our Occam's Razor hypothesis, then, is that the observed significant difference in both malarial parasitization and the efficacy of intervention between the dominant Fulani and co-resident ethnic groups in the Ouagadougou region of Burkina Faso is largely accounted for by factors of immunocultural condensation, particularly in view of the lower frequencies of classic malaria-resistance genes found in the Fulani.

Given their protective ICC, the Fulani simply may not need those classic genes.

It is not that the Fulani are not parasitized, or that the 'Fulani way' prevents disease, but that the population-level burdens of environment are modulated by historical development, and these are profoundly different for the (former) masters and the (former) slaves.

6.2.5 'Heterosexual AIDS' in Northern New Jersey

Studies by Skurnick et al. (1998) and Rohowsky-Kochan et al. (1998), under the general rubric of the Heterosexual Transmission Study (HATS), have examined 224 heterosexual couples discordant for HIV type 1 infection (one partner HIV infected) and for 78 HIV-concordant couples (both partners HIV-infected) to identify demographic and behavioral risk factors for HIV transmission. A large subset of this cohort was subsequently studied for differences in major histocompatibility complex (MHC)-encoded class I and class II antigens.

Couples were characterized by 'ethnicity' as ' Black, Non-Hispanic', 'White, non-Hispanic', and 'Hispanic'.

Skurnick et al. (1998) state

"In New Jersey, heterosexual transmission has played nearly as large a role in the AIDS epidemic as has injection drug use. Heterosexual contact was the category of transmission associated with the greatest increase in reported AIDS cases from 1994 to 1995. The severity of the epidemic and the frequency of heterosexual transmission in northern New Jersey motivated us... to evaluate the importance of behavioral and biological factors that facilitate or impede heterosexual transmission of HIV...

Risk factors that had significant bivariate associations with concordance were included in a multiple logistic regression model to evaluate their relative importance in their simultaneous effects on concordance... Ethnicity was the strongest correlate. Black and Hispanic couples were both more likely to be concordant [in HIV infection] than were whites or others."

This was no small effect. The odds ratio (OR) for concordance associated with 'Hispanic' ethnicity was 4.9(1.9-12.7, P=0.001), that for 'Black' a whopping 8.6(2.9-25.3, P=0.0001). The numbers in parenthesis are the 95% confidence limits and the associated P-value.

A principal conclusion of Skurnick et al. (1998) was that ethnicity may relate to genetic differences in susceptibility of the uninfected partner or infectiousness of the infected partner. That is, genetic factors entirely internal to the couples themselves primarily determine their concordant or discordant status.

The subsequent paper by Rohowsky-Kochan et al. (1998) examined the genetic hypothesis in more detail:

"Our results suggest that there may be different HLA alleles involved in the susceptibility and/or resistance to HIV infection in individuals of different ethnic backgrounds. It is possible that an as yet unidentified susceptibility/resistance genetic factor for HIV infection may be linked with different HLA alleles in different ethnic backgrounds...

The American Caucasians... are a very heterogeneous group comprised of a mixture of [identifiable] ethnic subpopulations...

[S]ignificant HLA associations with HIV resistance/susceptibility were detected in both Black and Hispanic cohorts but not in Caucasians... suggest[ing] that genetic factors may play a role in the finding of Skurnick et al. (1998) that Black and Hispanic heterosexual couples have a greater risk for HIV-1 concordance than Caucasian couples."

Again, alternative explanations consistent with the results are left unexplored in favor of a simplistic genetic reductionism.

European colonialism in the Americas parallels, in critical respects, that of the Fulani in Sub-Saharan Africa. 'Black' populations now speak English, and 'Hispanic' populations Spanish, and these terms efface tribal identity.

Although White ethnics can usually trace their past to some European homeland, African-Americans – 'Negroes', 'Blacks' – many of whom are, after two hundred years of sexual exploitation in slavery and under American Apartheid, more than a little 'White', usually cannot. Intermediate are the Hispanics in the US, who are, in spite of Spanish colonialism, more recognizably diverse. In Northern New Jersey they include self-identified Cubans, Puerto Ricans, Mexicans, Garafuna, Aymara, etc. etc., many of whom travel regularly to the homeland.

Northern New Jersey is, however, according to many studies (Massey and Denton, 1993; Acevedo-Garcia, 2000), one of the most heavily segregated regions of the US. Newark, the largest city in Northern New Jersey, in terms of what Massey and Denton (1993) call statistical measures of Unevenness, Isolation, Clustering, Centralization and Concentration, is even more segregated than nearby New York City, one of the world's most segregated cities. As Massey and Denton (1993) put it, comparing African-Americans and Hispanics,

"No other group in the contemporary United States comes close to this level of isolation within urban society. US Hispanics, for example, are also poor and disadvantaged; yet in no metropolitan area are they hypersegregated. Indeed, Hispanics are never highly segregated on more than three [of our five study factors] simultaneously... Despite their immigrant origins, Spanish language, and high poverty rates, Hispanics are considerably more integrated in US society than are blacks."

Given these circumstances, and taking malaria in Burkina Faso as a template, it seems evident that effects of the ICC in the context of the US system of Apartheid ensure that Caucasian couples would show fewer genetic markers of HIV than others, and that Blacks would show greater susceptibility to HIV transmission than Hispanics, and Blacks and Hispanics together, greater susceptibility than Caucasians.

6.2.6 Conclusions and speculations

At the individual level, as opposed to community scales of space, time and population, these matters are fairly well understood. Recent work by Kiecolt-Glaser and Glaser (1996, 1998, 2000), for example, has examined the effect of chronic stress on the efficacy of influenza, hepatitis-B, and pneumococcal pneumonia vaccine among elderly caregivers of dementia patients, and among medical students.

They found, for influenza, that the caregivers showed a poorer antibody response following vaccination relative to control subjects, as assessed by ELISA and hemagglutination inhibition. Caregivers also had lower levels of *in vivo* virus-specific-induced interleukin 2 levels and interleukin 1β. The data demonstrate that down-regulation of the immune response to influenza virus vaccination is associated with a chronic stressor in the elderly.

Similar effects were found among the elderly caregivers for response to pneumoccocal pneumonia vaccination, leading to the conclusion that chronic stress can inhibit the stability of the IgG antibody response to a bacterial vaccine.

Medical students who reported greater social support and lower anxiety and stress demonstrated a higher antibody response to HEP-B surface antigen at the end of the study period.

Glaser et al. (1998) conclude that the differences in antibody and T-cell responses to HEP-B and influenza virus vaccinations provide a demonstration of how stress may be able to alter both the cellular and humoral immune responses to vaccines and novel pathogens in both younger and older adults.

We reiterate that a vast body of animal model studies involving socially structured populations shows clear impacts of acute and chronic social and other stressors on immune competence (e.g., de Groot et al., 2001; Gryazeva et al., 2001; Stefanski et al, 2001). Elenkov and Chrousos (1999) in particular suggest that glucocorticoids and catecholamines, the end-products of the stress system at the individual level, might selectively suppress cellular immunity, Th1 phenotype, in favor of humoral response – again at the individual level.

We now suggest, however, that the essential role of culture in human biology takes matters considerably beyond such individual-level stress models, and into realms for which, to paraphrase Robert Boyd's aphorism, culture is as much a part of the human immune system as are T-cells. We have characterized the interaction between immune and sociocultural cognition as an immunocultural condensation, and use the concept to provide an Occam's Razor explanation of observed differences in patterns of malarial parasitization and response to intervention among co-resident ethnic groups in a section of Burkina Faso, and rates of heterosexual transmission of AIDS within different ethnic groups in Northern New Jersey.

The relation between the Fulani and the Rimaibe mirrors the relation between 'White' and 'Black' residents of the US. Thus we suspect that differences of ICC may play a large role in the health disparities evident between those groups, an effect which persists even in the face of adjustment for socioeconomic factors. This suggests the continuing burden of history – what organizational ecologists and evolutionary biologists have come to call path dependence – written upon the individual level ICC.

AIDS is a disease of marginalization and poverty, spreading along the structural flaws of a society like water through cracks in ice. Crossectional marginalization and deprivation are synergistic with longitudinal path dependent, historically driven, structures of ICC to define the ecology of the infection. This perspective, unlike current simplistic geneticism, does not reify 'race', but rather focuses on the central roles of culture, environment and development in the production of the 'quadruple helix' generating susceptibility to, and expression of, infection by pathogens. The analysis directly incorporates path dependence in a natural manner, making explicit the often-enduring effects of historical patterns of social, political, and economic exploitation. It goes well beyond cross-sectional socioeconomic status analyses.

To the degree that factors of ICC dominate a disease ecology, there is unlikely to be an effective, single, one-size-fits-all vaccine strategy. On the other hand, a more flexible attack which makes appropriate use of ICC mechanisms may enjoy a synergistic boost in effectiveness, at least among those who do not bear the burdens of history. For those who do bear the burdens, however, as the experiment with insecticide-treated curtains in Burkina Faso implies, circumstances may be difficult indeed.

Many of these matters should be directly testable, using immune system adaptations of Nisbett's (2001) experimental techniques.

In the next section, we add the effects of HIV's spatial economy on the virus' holistic evolution – a sociogeographic mode of farming pathogens.

6.3 Multiple Drug Resistant HIV in New York

6.3.1 Introduction

Human immunodeficiency virus (HIV) displays the strongest positive selection of any known organism. The virus, then, should be expected to successfully adapt to selection pressures generated by antiretrovirals, other microbicides, and vaccines. HIV can respond not only by developing multiple drug resistance, but also by significant alterations in life history strategy – increased virulence. Effective control of such a pathogen requires sophisticated multifactorial ecosystem intervention, including a return to traditional public health approaches aimed squarely at improving living and working conditions among the marginalized populations which are the keystones of pandemic infection.

Here we examine the likely impacts of a stunning counter strategy, the coevolutionary farming of the virus by a systematic program of forced displacement affecting poor African-Americans living in New York City's most heavily infected neighborhoods. The particular context is that the city is both the principal driving epicenter for the hierarchical spatial diffusion of emerging infections in the US and its economic partners, and is a central focus of HIV itself.

A conference held in 1993 by the Office of the High Commissioner for Human Rights of the United Nations characterized forced displacement in these terms (UNHCHR, 1993):

> "The practice of forced displacement involves involuntary removal of persons from their homes or land, directly or indirectly attributable to the State... The causes of forced evictions are very diverse. The practice can be carried out in connection with development and infrastructure projects... housing or land reclamation measures, prestigious international events, unrestrained land or housing speculation, housing renovation, urban redevelopment or city beautification initiatives, and mass relocation or resettlement programmes...
>
> The practice of forced displacement shares many characteristics with related phenomena such as population transfer, internal displacement of persons, forced removals during or as a result or object of armed conflict, 'ethnic cleansing', mass exodus, refugee movements, etc...."

Here we examine a decades-long process of forced displacement affecting African-Americans in New York City, with a special focus on the Harlem section of the Borough of Manhattan. We study the possible impacts of continuing displacement policies on the emerging scourge of multiply-drug resistant, or other evolutionarily transformed variants, of HIV.

A well-known report by Freeman and McCord (1990) examined excess mortality in Harlem, finding that, at the time, men in Bangladesh had a higher probability of survival after age 35 than men in Harlem. They noted, almost in passing, that Harlem's population had declined from 233,000 in 1960 to only 122,000 by 1980, with most of the population loss concentrated in the group living in substandard housing, much of it abandoned or partially occupied buildings. In that period the death rate from homicide increased from 25.3 to 90.8 per 100,000, with cirrhosis and homicide together accounting for some 33% of Harlem's excess deaths between 1979 and 1981. By 1990 AIDS became the most common cause of death for persons between 25 and 44 years of age in Harlem.

The policy-driven process inducing that depopulation has been described in some detail elsewhere (Wallace, 1990; R. Wallace and D. Wallace, 1997c; D. Wallace and R. Wallace, 1998, 2003; D. Wallace, 2001, and references therein). Withdrawal of essential housing-related municipal services, including fire extinguishment resources, from minority voting blocks in the 1970's triggered processes of large-scale contagious urban decay and forced migration involving a devastating synergism of fire, housing abandonment, and pathology (Wallace, 1988). The process was described by the New York State Assembly Republican Task Force on Urban Fire Protection (Task Force, 1978) as follows:

> "There is mounting evidence that the lack of fire protection which has plagued communities in the South Bronx, Central Harlem, Brownsville and Bushwick is assuming city-wide dimensions as it spreads to [other neighborhoods]... there are indications the City Planning Commission and other agencies condoned [fire service] reductions in the context of a 'planned shrinkage' policy... there is strong evidence that these actions have resulted in unwarranted loss of life and destruction of city neighborhoods..."

After examining the consequences Wallace (1990) wrote:

> "...[T]he... origins of public health and public order are much the same and deeply embedded in the security and stability of personal, domestic and community social networks and other institutions... [D]isruptions of such networks, from any cause, will express themselves in exacerbation of a nexus of behavior, including violence, substance abuse and general criminality. These in turn have the most severe implications for...[many pathologies including the] evolution and spread of AIDS."

The policy-driven displacement of population affecting Harlem between 1970 and 1980 created a massive de-facto refugee camp environment for emerging and re-emerging infection. By 1990 Harlem was an epicenter of both AIDS and tuberculosis, and of their interaction (D. Wallace and R. Wallace, 1998, 2003; R. Wallace and D. Wallace, 1997a; D. Wallace, 2001).

By 2005 Harlem and East Harlem rivaled the Gay center of New York City, Manhattan's Chelsea-Clinton neighborhood, in rates of HIV diagnoses per 100,000 population, respectively 132.4, and 108.2, vs 135.0. Age-adjusted death rates per 1000 persons with AIDS were, however, quite different: 31.9 and 32.6 vs. 11.4 (NYCDOH, 2006). This divergence represents not only a contrast in the effective availability of antretroviral drugs, but also obvious population differences in patterns of burden and affordance between African-Americans, other minorities, and middle-class Whites, in spite of similarly draconian pressures enforcing the social and spatial segregation of ethnic and sexual minorities in the United States (R.G. Wallace, 2003; Massey and Denton, 1992).

The planned shrinkage program, which exacerbated the spread of AIDS and tuberculosis (Wallace and Wallace, 1998), had, by 1990, set the stage for a subsequent round of displacement. The loss of economic, social, and political capital consequent on the induced contagious urban decay of the 1970's left Harlem without effective means of resisting 'gentrification' by majority populations, that is, the ongoing reduction of Harlem to a largely-White 'Central Park North'.

Newman and Wyly (2006) describe this as follows:

"Central Harlem received an influx of middle-class residents throughout the 1970s and 1980s but the changes during the late 1990s and early 2000s are different. Harlem's residents report a solid flow of SUV's (sports utility vehicles) of people driving through the neighborhood scouting for homes. One resident described the housing demand: 'People are coming up while you're on the street asking who owns the building. It's a daily thing'. The neighborhood also appeals to renters seeking livable space with manageable commutes. In less than 15 minutes, residents are whisked to midtown on a 2 or A train; in 30 minutes, they can reach jobs on Wall Street. A 20-minute cab ride gets you to LaGuardia Airport and every highway intersects with Harlem. Rents for floor-through apartments in brownstones are capturing $1,700 a month."

They conclude:

"According to neighborhood informants, many [of those displaced by the rise in rents] are moving out of the city to upstate New York, New Jersey and Long Island... Those who are forced to leave gentrifying neighborhoods are torn from rich local social networks of information and cooperation (the 'social capital' much beloved by policy-makers); they are thrown into an ever more competitive housing market shaped by increasingly difficult trade-offs between affordability, overcrowding and commuting accessibility to jobs and services. All of the pressures of gentrification are deeply enmeshed with broader inequalities of class, race and ethnicity, and gender... As affordable housing protections are dismantled in the current wave of neo-liberal policy-making, we are likely to see the end-game of gentrification as the last remaining barriers to complete neighborhood transformation are torn down... Low-income residents who manage to resist displacement may enjoy a few benefits from the changes brought by gentrification., but these bittersweet fruits are quickly rotting as the supports for low-income renters are steadily dismantled."

The forced displacement of New York City's African-American population to a suburban/exurban ring around New York City, much like the Black townships surrounding Capetown, can be expected to induce a new round of refugee camp behavioral syndromes which will further exacerbate the spread of HIV among African-Americans. At present African-Americans account for 15% of the US population, but constitute over half of new HIV infections.

Katrina-like dispersal of New York City's communities of color can be expected to fatally compromise:

[1] ongoing antiretroviral drug treatment of those already infected with HIV,

[2] the effectiveness of treating new cases with antiretrovirals, and

[3] virtually all possible infection prevention strategies.

This will, in all likelihood, markedly accelerate the development and spread of drug resistant viral strains.

Mathematical analysis of contagious process in a commuting field (Wallace et al., 1997, 1999; Wallace, 1999) suggests that, for national hierarchical diffusion, Metropolitan Regions are the systems of fundamental interest. From that perspective, a disease epicenter has much the same large-scale force of infection whether it is concentrated in the center, or dispersed around the periphery, of a particular large city: urban-suburban linkages are strong enough to create a functional equivalence (Wallace, 1997).

Given the powerful central role of New York City and its metropolitan region in the economic and political function of the American Empire, as we have come to know it, larger scale, that is, international, hierarchical diffusion of infection from it can be expected to occur. Recent elegant, and very disturbing, phylogeographic analysis by Gilbert et al. (2007) clearly confirms that the first wave of international spread of HIV was driven by incubation within the United States. Their work demonstrates convincingly that, for HIV-1 group M subtype B, the predominant variant in most countries outside of sub-Saharan Africa, while the virus had an initial transfer event from Africa to Haiti around 1966, the key to subsequent geographic diffusion was what then happened in the United States:

"...[A]ll..subtype B infections from across the world emanated from a single founder event linked to Haiti. This most likely occurred when the ancestral pandemic clade virus crossed from the Haitian community in the United States to the non-Haitian population there.... HIV-1 was circulating in one of the most medically sophisticated settings in the world for more than a decade before AIDS was recognized... [That is],[o]ur results suggest that HIV-1 circulated cryptically in the United states for $\approx$ 12 years before the recognition of AIDS in 1981."

The essential inference is not that 'AIDS originated in Haiti', but rather that HIV-1 group M subtype B became entrained into the US social and spatial system, strongly dominated by New York City and its metropolitan region, where it circulated for over a decade, and then spread hierarchically from the US to its trading partners. This is the pattern which may be expected for drug resistant or other evolutionarily transformed variants of the virus which are now incubating in the vast marginalized subpopulations of the United States.

A recent comprehensive review by Jones et al. (2008), which examined the pattern of emerging infections from 1940 to 2004, not just AIDS, broadly confirms this analysis. They found that two developed regions strongly dominated the expression of all new diseases in that period: the Northeast Corridor of the U.S, including Boston, New York, Philadelphia, Baltimore, and Washington D.C., each of which has a vast marginalized subpopulation, and the Greater London metro region.

We recapitulate something of the evolutionary biology of HIV and of the canonical pattern of hierarchical disease diffusion within the US, and end with the implications of African-American forced displacement for both national and international spread of evolutionarily transformed HIV.

6.3.2 Evolutionary biology of HIV

Affluent populations in the US have, at least in the short term, benefited greatly from the introduction of highly active antiretroviral therapy (HAART) against HIV. From 1995 to 1997, for example, HIV/AIDS deaths declined 63% in New York City, primarily among middle-class, and highly organized, Gay males (Chiasson et al., 1999). Declines in AIDS deaths have otherwise been quite heterogeneous, depending critically on both the economic resources and community stability of affected populations (e.g., R.G. Wallace, 2003).

At present, AIDS deaths in the US are, largely, another marker of long-standing patterns of racism and socioeconomic inequity (e.g., Wallace and McCarthy, 2006; R. Wallace et al., 2007). Those who have economic resources, or reside in stable communities not subject to various forms of redlining and/or de-facto ethnic cleansing, have effective access to HAART. Others, without resources, do not have such access.

HIV is, as indeed are most retroviruses, however, an evolution machine (Rambaut et al., 2004) which, at the individual level, almost always develops multiple drug resistance, resulting in overt AIDS and subsequent premature fatality. Such response to chemical pesticides, as has been the case with myriad other biological pests, is now becoming manifest at the population level. By 2001, in the US some 50% of patients receiving antiretroviral therapy were infected with viruses that express resistance to at least one of the available retroviral drugs, and transmission of drug-resistant strains is a growing concern (Clavel and Hance, 2004; Grant et al., 2002). Multiple drug resistant (MDR) HIV is, in fact, rapidly becoming the norm, and the virus may even develop a far more virulent life history strategy in response to the evolutionary challenges presented by HAART, its successor microbicide strategies, or planned vaccines (R.G. Wallace, 2004; Wallace and Wallace, 2004), a circumstance which may have already been observed (e.g., Simon et al., 2003).

The review by Rambaut et al. (2004) puts the matter thus:

> "HIV shows stronger positive selection than any other organism studied so far... [its viral] recombination rate... is one of the highest of all organisms... Within individual hosts, recombination interacts with selection and drift to produce complex population dynamics, and perhaps provide an efficient mechanism for the virus to escape from the accumulation of deleterious mutations or to jump between adaptive peaks. Specifically, recombination might accelerate progression to AIDS and provide and effective mechanism (coupled with mutation) to evade drug therapy, vaccine treatment or immune pressure... More worryingly, there is evidence that some drug-resistant mutants show a greater infectivity, and in some cases a higher replication rate, compared with viruses without drug resistant mutations."

R.G. Wallace (2004) finds that

> "...HAART may select for... an HIV with a semelparous life history and a precocious senescence... [which] may be embodied by an accelerated time to AIDS or related pathogenesis... Because infection survivorship is physiologically enmeshed with host survivorship the asymptomatic stage becomes under HAART a demographic shield against epidemiological intervention. The results appear to exemplify how pathogens use processes at one level of biological organization to defend themselves against impediments directed at them at another."

Above we have suggested that as population-level structured stress appears a fundamental part of the biology of many chronic infectious diseases including AIDS. This raises the possibility that simplistic individual-oriented magic bullet drug treatments, vaccines, and risk-reduction programs that do not address the fundamental living and working conditions which underlie disease ecology will fail to control many current epidemics. In addition, such reductionist interventions may go so far as to select for more holistic pathogens characterized by processes operating at multiple levels of biocultural organization.

MDR-HIV is already emerging in the very epicenters and epicenter populations where HIV itself first appeared (Clavel and Hance, 2004), since these were the first to benefit from HAART, and thus seems likely to follow diffusion patterns similar to those of the earlier stage of the AIDS epidemic. More general ET-HIV's can be expected to follow a similar pattern. We reconsider the initial period.

6.3.3 Hierarchical diffusion

Infectious disease is often seen as a marker for underlying urban structure. For example, Gould and Tornqvist (1971, p. 160) write:

"As the urban lattice hardens, and the links between the major centers strengthen, the dominant process is apt to change from a [spatially] contagious to a hierarchical one.

We have few examples of this dramatic change in innovation diffusion, but one particularly striking one comes from the early history of the United States (Pyle, 1969). The disease cholera is hardly an innovation we would like to spread around, but it does form a useful geographical tracer in a spatial system, rather like a radioactive isotope for many systems studied by the biological sciences. The first great epidemic struck in 1832 at New York and Montreal, and then diffused slowly along the river systems of the Ohio and Great Lakes. A graphical plot of the time the disease was first reported against distance shows a clear distance effect, indicating that basically processes of spatial contagion were operating. A plot of time against city size shows no relationship whatsoever. However, by 1849, the rudimentary urban hierarchy of the United States was just beginning to emerge. The second epidemic struck at New York and New Orleans in the south, and a plot of first reporting times against city size, indicates that a hierarchical effect was beginning to structure innovation flows at this time. Finally, in 1865, when the third epidemic struck, the railways were already strengthening the structure of America's urban space. The disease jumped rapidly down the urban hierarchy, and a plot of reporting time against city size shows that a very clear hierarchical process was at work."

The first stages of the AIDS pandemic in the US provide a modern example. The cover of Gould's 1993 book *The Slow Plague*, with more detail in Gould (1999), presents a time sequence of maps showing the number of AIDS cases in the US on a logarithmic scale. Cases first appear in the largest US port cities: New York, Los Angeles, San Francisco, Miami and Washington DC. Subsequent spread is by hierarchical hopscotch to smaller urban centers, followed by a spatially contagious winestain-on-a-tablecloth diffusion from city center into the surrounding suburban counties.

Figure 6.1, from Wallace et al. (1999), gives a detailed analytic treatment of the hierarchical hopscotch. Using multivariate analysis of covariance, it shows the log of the number of AIDS cases in each of the 25 largest US metropolitan regions for two periods, [1] through April, 1991 and [2] from April 1991 through June 1995, as functions of a composite index defined in terms of a region's local pattern of susceptibility and its position in the US urban hierarchy. The local indices are (i) the log of the number of violent crimes in the region for 1991, and (ii) an index of 'rust belt' deindustrialization, the log of the ratio of manufacturing employment in 1987 to that in 1972. The global index, of position on the US urban hierarchy, is the log of the probability of contact with the New York City metro region, the nation's largest, determined from a county-by-county analysis of migration carried out by the US Census for the period 1985-1990.

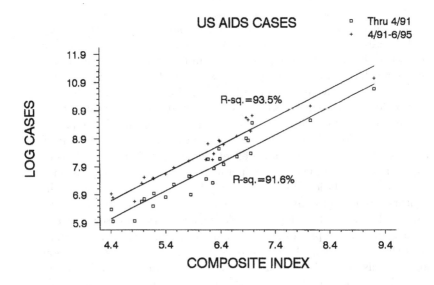

Fig. 6.1. Log number of AIDS cases in the 25 largest US metropolitan regions, through 4/91 and 4/91-6/95. The composite index is X = .764 Log(USVC91) + .827 Log(USME87/USME72) + .299 Log(Prob. NY). USVC91 is the number of violent crimes, USMEnm the number of manufacturing jobs in year nm, and (Prob. NY) the probability of contact with New York City according to Census migration data for 1985-1990. Applying multivariate analysis of covariance, the two lines are parallel with different intercepts: The second is obtained simply by raising the first. The New York Metropolitan Region (NYMR) is at the upper right of the graph. The pattern is consistent with the assumption that the NYMR drives the hierarchical diffusion of HIV nationally. This suggests a powerful, coherent, national-scale, spatiotemporal hierarchical diffusion strongly linking marginalized communities in the NYMR, the Apartheid policies which marginalize them, and the epidemic outbreak within them, to the rest of the country.

Locally, high levels of violence and industrial displacement represent bust-town and boom-town social dynamics leading to the loosening of social control. Nationally, the probability of contact with New York represents inverse socio-spatial distance from the principal epicenter of the US AIDS epidemic. Multivariate analysis of covariance finds the lines for the two time periods are parallel and each accounts for over 90 % of the variance in the dependent variate. Thus later times are obtained from the earlier simply by raising the graph in parallel. This means that processes within the New York Metropolitan Region, the upper right point of the graph, drive the national hierarchical diffusion of AIDS during this time span, the pre-HAART period of AIDS spread. We take this as representing a propagating, spatio-temporally coherent epidemic process which has linked disparate, marginalized 'core group' neighborhoods of Gay males, intravenous drug users, and ethnic minority populations across time and space with the rest of the urbanized US, ultimately placing some 3/4 of the nation's population at increasing risk: as go the New York Metropolitan Region's segregated communities, so goes the Nation. See Wallace, Wallace et. al (1999) for details, and Abler et al. (1971) or Gould (1993, 1999) for more general background.

The results of Gilbert et al. (2007) prove conclusively that a similar pattern affects the hierarchical spread of HIV variants from the US to its principal trading partners. The US, and by inference its dominant conurbation, the New York Metropolitan Region, served as the primary source for diffusion of HIV-1 group M subtype B to the industrialized world, and will almost certainly continue to do so for its evolutionary transformed variants.

The social and geographic spread of infectious disease within a polity is constrained by, and must be consistent with, an underlying sociogeography in which segregated and oppressed subgroups traditionally constitute ecological keystone populations. Acutely marginalized communities within and surrounding the largest cities are particularly central.

In contrast to the conclusions of Jones et al., (2008), who inferred a 'mismatch' in global resources currently allocated to emerging infectious disease surveillance and control, with developed countries receiving, in their view, an unneeded preponderance of resources, we see a two-fold, synergistic process affecting diffusion of HIV and other emerging infections, which may be summarized as:

[1] All roads lead to Rome.

[2] All roads lead from Rome.

That is, at first, de-facto colonial exploitation of economically peripheral zones by the American Empire and its more developed client states, creates circumstances ripe for the emergence of new infectious disease by a variety of mechanisms. These pathogens are then entrained by economically determined travel patterns into the largest urban centers of the US and its allies (in particular the Northeast Corridor of the US, and the London Metro Region), which serve as the New Rome of the current imperial system.

Second, the domestic version of the US colonial enterprise, in particular the fatal legacy of slavery, which developed into the present system of American Apartheid (*sensu* Massey and Denton, 1992), has created vast marginalized populations within the nation's largest metropolitan regions, particularly the Northeast Corridor. Newly-imported emerging infections then incubate largely unnoticed within these huge permanent de-facto refugee camps, and subsequently blow back down the US urban hierarchy, and across to its more developed client states, in particular the European Union.

There are several implications of this model for HIV. Most simply, the find-the-cure 'Treatment Culture' which has dominated both official and non-governmental organizations' AIDS policy in the US for some time, and particularly since the development of HAART, is ending as HIV evolves resistance to drug regimens or alters its life history (R.G. Wallace, 2004; Simon et al., 2003; Wallace and Wallace, 2004). Possible vaccines seem similarly challenged by HIV's protean evolutionary nature, which allows the virus rapidly evolve out from under immune suppression. Indeed, a consensus is now emerging that the twenty-year search for a vaccine against HIV has made no progress, with little hope for future results (Baltimore, 2008).

As the AIDS Treatment and Vaccine subcultures disintegrate under the relentless pressure of pathogen evolution new social organizations must emerge to confront the disease. Traditional public health approaches, which address underlying structural factors responsible for disease incubation and spread at the population level – primarily the power relations between groups – have largely been abandoned in the US for reasons of political expediency. The field is now dominated by a kind of neo-liberal, rightist intellectual analog to the Stalinist sophistry of the fallen Soviet Union, effectively a Center-Right Lysenkoism strongly driven by funding biases.

Controlling MDR, vaccine-resistant and other ET-HIV's will require resurrection of traditional public health, but this will be difficult because, in the US, so much of the discipline's history has been lost in favor of the blame-the-victim, medicalized, and individual-oriented perspectives now popular with the current crop of major AIDS funding agencies and their client organizations. Many resulting projects are characterized as 'fundable trivialities' or 'planting a tree in a desert' even by foundation staff administering financial support.

ET-HIV's are now poised to spread from the main US AIDS epicenters, particularly the New York Metropolitan Region, following much the same patterns as the pre-HAART pandemic. In contrast, evidence exists that, for at least one more egalitarian social system – Amsterdam – there is a declining trend in transmission of drug-resistant HIV (Bezemer et al., 2004).

What is also clear, at least in the New York Metropolitan Area which drives the hierarchical diffusion of all emerging infection (Gould, 1993, 1999), and thus a keystone in the international spread of HIV variants from the US to its trading partners (Gilbert et al., 2007; Jones, et al., 2008), is that rebuilding of housing lost to prior policies of ethnic cleansing and stabilizing remaining low income housing, is necessary for both regional and national control of ET-HIV's. The current gentrification of Central Harlem and other communities of color, given the region's recent history, will simply further tighten the contact probability field which links metropolitan counties (e.g. Wallace and McCarthy, 2007 and references therein), and thus hasten the diffusion of the new virus within the New York Metropolitan Region. As a result of its dominance and sociogeographic centrality in humanity's ecosystem, New York HIV will be catapulted throughout the United States and the rest of the world.

In sum, displacing African-American and other poor populations from Harlem and East Harlem into the lower-rent suburban periphery will create refugee camp conditions in the outlying New York Metropolitan Region for the development and international spread of evolutionarily-transformed HIVs, including multiple-drug-resistant strains of the virus. Urban-suburban linkages are strong in and near New York City, and can be expected to link periphery and center in a new, virulent ecology of ET-HIV which will then diffuse down the urban hierarchy, then from city center to suburban ring, effectively placing at increased risk some 3/4 of the US population. Subsequent spread of ET-HIV's to other industrialized countries will likely follow the pattern uncovered by Gilbert et al. (2007).

At present, with condominiums presently selling for an average of well over US $1.2 million each in Manhattan, assembling a package of only a dozen or so 100 unit apartment houses in Harlem represents a potential profit of nearly US $1 billion. For a billion dollars most developers, and their clients among public officials and the leaders of non-governmental organizations, are unlikely to give much thought to the national and international diffusion of multiple-drug-resistant HIV. Pushing poor people out offers too much opportunity for profit. The results seem likely to establish the New York Metropolitan Region as the global source of MDR-HIV in much the same manner that China's Guangdong Province has, for deep structural reasons, become the epicenter for worldwide transmission of highly pathogenic influenza A H5N1. the avian influenza virus.

6.4 Avian influenza

6.4.1 Panic in the city

Hong Kong, March 1997. An outbreak of deadly avian influenza sweeps through poultry on two farms (Davis, 2005; Greger, 2006). The outbreak subsides, but two months later a three-year-old boy dies of the same strain, identified as a highly pathogenic version of influenza A (H5N1). Officials are shocked. This appears the first time such a strain has jumped the species barrier and infected a human. Shocking too, the outbreak proves persistent. In November a six-year-old is infected, recovering. Two weeks later, a teenager and two adults are infected. Two of the three die. Fourteen additional infections rapidly follow.

The deaths spur panic in the city and, with the onset of the regular flu season, send many patients to the hospital worried their symptoms might be those of the new flu. By mid-December poultry begin to die in droves in the city's markets and it now seems most humans infected had handled birds. Hong

Kong acts decisively on that information. Authorities order the destruction of all of Hong Kong's 1.5 million poultry and block new imports from Guangdong, the mainland province across the Shenzhen River from which some of the infected birds had been transported. Despite another human death in January, the outbreak is broken.

The poultry infected with this version of the virus suffer more than the gastrointestinal condition typical of avian influenza. The clinical manifestations include swelling of the wattles and infraorbital sinuses, congestion and blood spots on the skin of the hocks and shanks, and a blue discoloration of the comb and legs (Yuen and Wong, 2005). The latter is characteristic of the cyanosis and oxygen deprivation suffered by many human victims of the 1918 pandemic. Internally, infected poultry are marked by lesions and hemorrhaging in the intestinal tract and the trachea, with blood discharge from the beak and cloaca. Many birds also suffer infection in other organs, including the liver, spleen, kidney, and the brain, the last infection leading to ataxia and convulsion.

Most worrisome for human health is this strain's capacity for broad xenospecific transmission. The Hong Kong outbreak, first alerting the world to H5N1, infected humans with an influenza much more pathogenic than the relatively mild infections of other avian outbreaks that have intermittently crossed over into human populations. These patients presented with high fever, later developing some combination of acute pneumonia, influenza-like illness, upper respiratory infections, conjunctivitis, pharyngitis, and a gastrointestinal syndrome that included diarrhea, vomiting, vomiting blood, and intestinal pain (Bridges et al., 2000; de Jong et al. 2006). Patients also suffered multiple-organ dysfunction, including that of the liver, kidney, and bone marrow. The respiratory attacks involved extensive infiltration of both lungs, diffuse consolidation of multiple infected loci, and lung collapse.

If much of H5N1's morbidity is distressing, its associated mortality is alarming. Once infected, the lungs' vasculature becomes porous and fibrinogen-a protein involved in blood clotting-leaks into the lungs (de Jong et al., 2006). The resulting fibroblast exudates clog the lungs' alveolar sacs, where gas exchange takes place, and an acute respiratory disease syndrome results. In a desperate effort to save its charge, the immune system recruits such a storm of cytokines that the lungs suffer oedema. In effect, patients drown in their own fluid only days after infection.

After Hong Kong, H5N1 slipped somewhat underground with local outbreaks largely limited to birds in southern China. During this time the virus underwent the first of a series of reassortment events, in which several genomic segments were replaced with those from other serotypes, reemerging as a human infection in Hong Kong in 2002 (Li et al., 2004; Webster et al., 2006). The following year H5N1 again reemerged, this time with a vengeance. The Z genotype that surfaced as the dominant recombinant spread across China, into Vietnam, Thailand, Indonesia, Cambodia, Laos, Korea, Japan, and Malaysia. Two additional strains would subsequently materialize. Since

2005 the Qinghai-like strain has spread across Eurasia, as far west as England, and into Africa (Salzberg et al. 2007). The Fujian-like strain, emerging from its eponymous southern China province, has spread regionally across Southeast Asia and, more recently, into Korea and Japan (Smith et al., 2006).

Since 2003 H5N1 has infected 387 humans, killing 245 (WHO, September, 2008). Most of these infections have been poultry related; often the children of small farmers playing with a favorite bird. But an increasing number of documented cases of human-to-human transmission have accumulated-in Hong Kong, Thailand, Vietnam, Indonesia, Egypt, China, Turkey, Iraq, India, and Pakistan (Kandan et al., 2006; Yang et al., 2007). The short chains of transmission have largely consisted of relatives living with or tending a patient. The worry, well publicized, is that H5N1 will improve upon these first infections, evolving a human-to-human phenotype that ignites a worldwide pandemic.

The geographic diffusion of the virus is intimately related to the emergence of such a phenotype. As are other pathogens, H5N1 is finding the regions of the world where animal health surveillance remains underdeveloped or degraded by national structural adjustment programs associated with international loans (Rweyemamu et al., 2000). There is now too a greater integration of aquaculture and horticulture, a burgeoning live-bird market system, and widespread proximity to backyard fowl (Gilbert et al., 2007; Cristalli and Capua, 2007). Rural landscapes of many of the poorest countries are now characterized by unregulated agribusiness pressed against periurban slums (Guldin, 1993; Fasina et al., 2007). Unchecked transmission in vulnerable areas increases the genetic variation with which H5N1 can evolve human-specific characteristics. In spreading over three continents fast-evolving H5N1 also contacts an increasing variety of socioecological environments, including locale-specific combinations of prevalent host types, modes of poultry farming, and animal health measures.

In this way, by a type of escalating demic selection, H5N1 can better explore its evolutionary options (Wallace and Wallace, 2003). A series of fit variants, each more transmissible than the next, can evolve in response to local conditions and subsequently spread. The Z reassortant, the Qinghai-like strain, and the Fujian-like strain all outcompeted other local H5N1 strains to emerge to regional and, for the Qinghai-like strain, continental dominance. The more genetic and phenotypic variation produced across geographic space, the more compressed the time until a human infection evolves.

6.4.2 Farming deadly influenza

Despite its impacts epidemiological and psychological, Hong Kong's H5N1 represented no first outbreak of avian influenza. In fact, within the United States alone, where the southern Chinese H5N1 has yet to reach, there have been a series of outbreaks the past decade. These outbreaks were typically low pathogenic, causing lesser damage to poultry. There was, however, an outbreak of highly pathogenic H5N2 in Texas in 2002. A low pathogenic H5N2 outbreak in California, beginning in farms outside of San Diego, evolved greater virulence as it spread through California's Central Valley (Davis, 2005). Another outbreak worthy of note is that of a low pathogenic strain of H5N1 in Michigan in 2002. H5N1, then, has already invaded the United States in a less deadly form, telling us that the molecular identity of a strain is insufficient for defining the danger of any single outbreak. Low and high pathogenic strains must be distinguished otherwise. Some mechanism must transform low pathogenic strains into more virulent ones (and, we should hope, back again).

The nastiness of the southern Chinese H5N1 may be in part due to an antigenic shift to which we presently have no immunity. Humans have this past century been infected almost exclusively by H1, H2, and H3 strains to which we have developed antibody memory. When many of us are confronted by another strain of these same types we can slow down the infection. We have partial immunity at the individual level and herd immunity at the population level. Since we have never been exposed to H5 infections *en masse* we have nothing to slow down infection within each person and nothing to keep it damped down across the population. What cannot be slowed down arrives earlier. It is likely then that, as was the case for the 1957 and 1968 pandemics, the main wave of the next new human-to-human influenza will sweep the planet earlier than the typical seasonal flu, perhaps even as early as August some terrible year in the near future (Cliff et al. 1986).

But how are we to account for an increase in virulence *within* a particular flu subtype? Recall the low pathogenic strain of H5N1 in Michigan. Another explanation leans on a large modeling literature (reviewed by Dieckmann et al., 2002; Ebert and Bull, 2008) that hypothesizes a relationship between the rate of transmission and the evolution of virulence, the amount of damage a strain causes its host. Simply put, to start, there is a cap on pathogen virulence. Pathogens must avoid evolving the capacity to incur such damage to their hosts that they are unable to transmit themselves. If a pathogen kills its host before it infects the next host it destroys its own chain of transmission. But what happens when the pathogen 'knows' that the next host is coming along much sooner? The pathogen can get away with being virulent because it can successfully infect the next susceptible in the chain before it kills its host. The faster the transmission rate, the lower the cost of virulence.

A key to the evolution of virulence is the supply of susceptibles (Lipsitch and Nowak, 1995). As long as there are enough susceptibles to infect, a virulent phenotype can work as an evolutionary strategy. When the supply runs out it does not matter what virulence a pathogen has evolved. Time is no longer on the particular strain's side. A failed supply of susceptibles, drained by high mortality or rebound immunity, forces all influenza epidemics to ultimately burn out at some point. That's cold comfort, of course, if millions of people are left dead in a pandemic's wake.

What caused southern China's H5N1 to evolve its breathtaking virulence? The circumstantial evidence points overwhelmingly to factory farming (Shortridge, 2003; FAO, 2004; Greger 2006). Growing genetic monocultures removes whatever immune firebreaks may be available to slow down transmission (Garrett and Cox, 2008). Larger poultry population sizes and densities facilitate greater transmission. Such crowded conditions depress immune response. High turnover, a part of any industrial production, provides a continually renewed supply of susceptibles, the fuel for the evolution of virulence.

There are additional pressures on influenza virulence on such farms. As soon as industrial poultry reach the right bulk they are killed. Think 'No Factory for Old Chickens,' with Javier Bardem as the plant manager. Resident influenza infections must reach their transmission threshold quickly in any given bird, before the chicken or duck or goose is sacrificed. The quicker viruses are produced, the greater the damage to the chicken. Increasing age-specific mortality in factory chickens should select for greater virulence. With innovations in production the age at which chickens are processed has been reduced from 60 days to 40 days (Striffler, 2005), increasing pressure on viruses to reach their transmission threshold – and virulence load – that much faster.

Along with hosting experiments in mounting virulence, industrial production has also increased the diversity of human-friendly influenza. Over the past 15 years an unprecedented variety of influenzas capable of infecting humans has emerged across the global archipelago of factory farms. Along with H5N1 there are H7N1, H7N3, H7N7, H9N2, in all likelihood H5N2, and perhaps even some of the H6 serotypes (WHO 2005, Puzelli et al. 2005, Meyers et al., 2007; Ogata et al., 2008). Something of a positive loop appears to have emerged in kind: the very efforts pursued to control pathogenic avian influenza may in passing increase viral diversification. In late 2006, virologist Guan Yi and his colleagues at the University of Hong Kong identified the previously uncharacterized Fujian-like H5N1 lineage (Smith et al., 2006). The team ascribed the emergence of the strain as a viral evolutionary reaction to the Chinese government's campaign to vaccinate poultry. As in the case of other influenza serotypes (Suarez et al., 2006; Escorcia et al., 2008), the virus appeared to evolve out from underneath the pressure of vaccine coverage.

Factory farms provide what seems to be an ideal environment for the evolution of a variety of virulent influenzas. And that seems to be a cost agribusiness is willing to incur for the cheaper manufacture of its product.

6.4.3 The political virology of offshore farming

In Israel recently researchers selected for a lineage of featherless chickens (Yaron et al., 2004). The birds look like walking groceries, ready to hop up into the meats freezer in aisle 6 of your local supermarket. These chickens, able to survive in warm climes alone, were developed in the interests of the producer, not the consumer. Consumers have long avoided plucking feathers. That's typically done at the factory. A featherless poultry will allow producers, on the other hand, to scratch off plucking feathers from production. The bald bird offers the anatomical equivalent of the factory epidemiology agribusiness is imposing on poultry-generating artificial chicken ecologies that could never persist in nature because of the epidemiological costs they incur, but that allow more poultry to be processed faster. The resulting costs are shifted to consumers and taxpayers alike.

The lengths to which agribusiness has changed poultry production are remarkable, including, more recently, in the present avian influenza zone. Southern China serves as a regional incubator for new methods in poultry breeding (Luo et al., 2003); Sun et al., (2007), for instance, describe a Guangdong program in which geese were exposed to a counter-seasonal lighting schedule that induced out-of-season egg-laying. The innovation helped double profits for local goose production and expanded the market, and Chinese appetite, for goose meat. The resulting market advantages forced smaller farms out of business and led to a consolidation of the province's agribusiness. The structural shift marks a perverse turn back toward the farm collectivization the Chinese government abandoned in 1980, this time, though, under the control of far fewer hands.

Karl Marx (1867/1990) traced many of the fundamentals of such efforts at commodification. In the first chapter of the first volume of *Capital* Marx wrote that human-made objects have multiple characteristics. They have use value – a hammer can be used to beat down nails. In all human economies objects also have an exchange value-how many other objects (say, screwdrivers) for which a hammer can be exchanged. A capitalist economy adds a third characteristic, turning objects into commodities. Surplus value is that part of the object's worth that accrues to capitalists as profit. Marx's contribution was showing that capitalists expropriate the surplus value by taking it out of the value that workers added to the commodity when they make it, usually by paying workers lower wages or increasing worker productivity, paying them the same or less for more work.

In our efforts to better understand how influenza evolves we need only address here Marx's point that capitalists produce commodities not because commodities are useful – have use value – but because they accrue surplus value, to capitalists the most important characteristic of the object. Changing the color or style of a hammer to attract more consumers may not seem such a big deal, but for other objects changes in use value can have far-reaching, even dangerous, consequences. In this case, agribusiness has changed its commodity – living, breathing organisms-to maximize productivity. But what does it mean to change the use value of the creatures we eat? What happens when changing use value turns our poultry into plague carriers? Does out-of-season goose production, for instance, allow influenza strains to avoid seasonal extirpation, typically a natural interruption in the evolution of virulence? Are the resulting profits defensible at such a cost to the rest of us?

Mass commodification of poultry emerged in what is now called the 'Livestock Revolution.' Before the great shift, poultry was largely a backyard operation. In Boyd and Watts' (1997) map of poultry across the United States in 1929, each dot represents 50,000 chickens. We see wide dispersion across the country-300 million poultry total at an average flock size of only 70 chickens. The production filiere of that era shows local hatcheries sold eggs to backyard poultry producers and independent farmers, who in turn contracted independent truckers to bring live poultry to city markets.

That changed after WWII. Tyson, Holly Farms, Perdue, and other companies vertically integrated the broiler filiere, buying up other local producers and putting all nodes of production under one company's roof (Manning and Baines, 2004; Striffler, 2005). Boyd and Watts show by 1992 US poultry production is largely concentrated in the South and parts of a few other states. Each dot now represents 1 million broilers, 6 billion in total, with average flock size of 30,000 birds.

By the 1970s, the new production model was so successful it was producing more poultry than people typically ate. How many roasted chickens were families prepared to eat a week? With the assistance of food science and marketing the poultry industry repackaged chicken in a mind-boggling array of new products, including chicken nuggets, strips of chicken for salads, and cat food. Multiple markets were developed large enough to absorb the production.

Industrial poultry also spread geographically. With production widespread, world poultry meat increased from 13 million tons in the late 1960s to about 62 million by the late 1990s, with the greatest future growth projected in Asia (FAO, 2003). In the 1970s Asian-based companies such as Charoen Pokphand set up vertical filieres in Thailand and, soon after, elsewhere in the region. Indeed, CP was the very first foreign company allowed to set up production in Guangdong under Deng Xiaoping's economic reforms. China has since hosted a veritable explosion in annual chickens and ducks produced (Gilbert et al., 2007). Increases in poultry have also occurred throughout Southeast Asia, though not nearly at the magnitude of China.

According to geographer David Burch (2005), the shift in the geography of poultry production has some very interesting consequences. Yes, agribusinesses are moving company operations to the Global South to take advantage of cheap labor, cheap land, weak regulation, and domestic production hobbled in favor of heavily subsidized agro-exporting (Manning and Baines, 2004; McMichael, 2006). But companies are also engaging in sophisticated corporate strategy. Agribusinesses are spreading their production line across much of the world. For example, the CP Group, now the world's fourth largest poultry producer, has poultry facilities in Turkey, China, Malaysia, Indonesia, and the US. It has feed operations across India, China, Indonesia, and Vietnam. It owns a variety of fast food chain restaurants throughout Southeast Asia.

Such rearrangements falsify the widely promulgated assumption that the market corrects corporate inefficiencies. On the contrary, vertical multinationalism cushions companies from the consequences of their own mistakes. First, multinationals producing at scale can price unprotected local companies out of business – the Wal-Mart effect. Consumers have nowhere else to go to punish subsequent corporate blunders. Secondly, by threatening to move operations abroad multinationals can control local labor markets; hobbling unions, blocking organization drives, and setting wages and working conditions. Unions are an important check on production practices that affect not only workers and consumers, but both directly and by proxy the animals involved in production. Thirdly, vertical agribusiness acts as both poultry supplier and retailer. The CP Group, for instance, owns a number of fast-food chains in a number of countries selling, what else, CP chicken. In short, fewer independent chains exist to play suppliers off each other by demanding the food be prepared in a way healthy to animals and humans alike.

In operating factories across multiple countries multinationals can hedge their bets in a variant of David Harvey's (1982/2006) spatial fix. The CP Group operates joint-venture poultry facilities across China, producing 600 million of China's 2.2 billion chickens annually sold (Burch, 2005). When an outbreak of avian influenza occurred in a farm operated by the CP Group in Heilongjiang Province, Japan banned poultry from China. CP factories in Thailand were able to take up the slack and increase exports to Japan. With the price per poultry ton increasing in the wake of an avian influenza crisis it helped create, the CP Group grossed greater profits. A supply chain arrayed across multiple countries increases the risk of avian influenza spread even as it allows some companies the means by which to compensate for the resulting interruptions in business (Sanders, 1999; Manning et al., 2007).

To protect the interests of agribusiness even as its operations struggle or fail, multinationals also fund politicians or field their own candidates. Thaksin Shinawatra, the Prime Minister of Thailand during the country's first avian influenza outbreaks, came to power on the backs of the telecommunications and livestock industries. Shinawatra played a prime role in blocking Thai efforts to control avian influenza. As Mike Davis (2005) describes it, when outbreaks began in Thailand, corporate chicken-processing plants accelerated production. According to trade unionists processing increased at one factory from 90,000 to 130,000 poultry daily, even as it was obvious many of the chickens were sick. As word got out about the illness, Thailand's Deputy Minister of Agriculture made vague allusions to an 'avian cholera' and Shinawatra and his ministers publicly ate chicken in a show of confidence.

It later emerged that the CP Group and other large producers were colluding with government officials to pay off contract farmers to keep quiet about their infected flocks. In turn, livestock officials secretly provided corporate farmers vaccines. Independent farmers, on the other hand, were kept in the dark about the epidemic, and they and their flocks suffered for it. Once the cover-up was blown open, the Thai government called for a complete modernization of the industry, including requiring all open-air flocks exposed to migratory birds be culled in favor of new biosecure buildings only wealthier farmers could afford.

Attempts to proactively change poultry production in the interests of stopping avian influenza can be met with severe resistance by governments beholden to corporate sponsors. In effect, H5N1, by virtue of its association with agribusiness, has some of the most powerful representatives available defending its interests in the halls of government. The very biology of avian influenza is enmeshed with the political economy of the business of food.

If multinational agribusinesses can parlay the geography of production into huge profits, regardless of the outbreaks that may accrue, who pays the costs? The costs of factory farms have long been externalized. As Peter Singer (2005) explains, the state has been forced to pick up the tab for the problems these factories cause; among them, health problems for its workers, pollution released into the surrounding land, food poisoning, and damage to transportation infrastructure. A breach in a poultry lagoon, releasing a pool of poultry shit into a Cape Fear tributary that causes a massive fish kill, is left to local governments to clean up.

With the specter of avian influenza the state is again prepared to pick up the bill so that farm factories can continue to operate without interruption, this time in the face of a worldwide pandemic agribusiness helped cause in the first place. The economics are startling. The world's governments are prepared to subsidize agribusiness billions upon billions for damage control in the form of animal and human vaccines, Tamiflu, and body bags. Along with the lives of millions of people, the establishment appears willing to gamble much of the world's economic productivity, which stands to suffer catastrophically if a pandemic were to erupt.

6.4.4 Why Guangdong? Why 1997?

In reorganizing its poultry industry under the American model of vertically integrated farming, Chinese farming helped accelerate a phase change in influenza ecology, selecting for strains of greater virulence, wider host range, and greater diversity. For decades a variety of influenza subtypes have been discovered emanating from southern China, Guangdong included (Chang, 1969; Shortridge and Stuart-Harris, 1982; Xu et al., 2007; Cheung et al., 2007). In the early 1980s, with poultry intensification under way, University of Hong Kong microbiologist Kennedy Shortridge (1982) identified 46 of the 108 different possible combinations of hemagglutinin and neuraminidase subtypes circulating worldwide at that time in a single Hong Kong poultry factory.

Shortridge detailed the likely reasons southern China would serve as ground zero for the next influenza pandemic:

[1] Southern China hosts mass production of ducks on innumerable ponds, facilitating fecal-oral transmission of multiple influenza subtypes.

[2] The greater mix of influenza serotypes in southern China increases the possibility the correct combination of gene segments would arise by genetic reassortment, selecting for a newly emergent human strain.

[3] Influenza circulates year-round there, surviving the interepidemic period by transmitting by the fecal-oral mode of infection.

[4] The proximity of human habitation in southern China provides an ideal interface across which a human-specific strain may emerge.

The conditions Shortridge outlined twenty-five years ago have since only intensified with China's liberalizing economy. Millions of people have moved into Guangdong the past decade, a part of one of the greatest migration events in human history, from rural China into cities of the coastal provinces (Fan, 2005). Shenzhen, one of Guangdong's Special Economic Zones for open trade, grew from a small city of 337,000 in 1979 to a metropolis of 8.5 million by 2006. As discussed earlier, concomitant changes in agricultural technology and ownership structure have put hundreds of millions more poultry into production (Luo et al., 2003; Burch, 2005; Sun et al., 2007). Poultry output increased in China from 1.6 million tons in 1985 to nearly 13 million tons by 2000.

As Mike Davis (2005) summarizes it, by the onset of pathogenic H5N1, only the latest pathogen to emerge under such socioecological conditions,

> "[S]everal subtypes of influenza were traveling on the path toward pandemic potential. The industrialization of south China, perhaps, had altered crucial parameters in the already very complex ecological system, exponentially expanding the surface area of contact between avian and nonavian influenzas. As the rate of interspecies transmission of influenza accelerated, so too did the evolution of protopandemic strains."

Pathogenic H5N1's hemagglutinin protein was first identified by Chinese scientists from a 1996 outbreak on a goose farm in Guangdong (Tang et al., 1998). News reports during the initial H5N1 outbreak in Hong Kong detailed local health officials' decision to ban poultry imports from Guangdong from where several batches of infected chickens originated (Kang-Chung, 1997). Phylogeographic analyses of H5N1's genetic code have pointed to Guangdong's role in the emergence of the first and subsequent strains of pathogenic H5N1 (Wallace et al., 2007). Scientists from Guangdong's own South China Agricultural University contributed to a 2005 report showing that a new H5N1 genotype arose in western Guangdong in 2003-4 (Wan et al., 2005).

Subsequent work has complicated the picture. With additional H5N1 samples from around southern China, Wang et al. (2008) showed virus from the first outbreaks in Thailand, Vietnam and Malaysia appeared most related to isolates from Yunnan, another southern Chinese province. Indonesia's outbreaks were likely seeded by strains isolated from the province of Hunan. These are important results, showing the complexity of influenza's landscape. At the same time they need not absolve Guangdong. Even if some H5N1 strains emerged elsewhere in the region, Guangdong's socioeconomic centrality may have acted as an epidemiological attractant, drawing in novel poultry trade-borne strains from around southern China before dispersing them again back out across China and beyond.

Mukhtar et al. (2007) meanwhile traced the origins of the genomic segments from the original 1996 outbreak in Guangdong. Most of the internal proteins (encoding for proteins other than surface proteins hemagglutinin and neuraminidase) appeared phylogenetically closest to those of H3N8 and H7N1 isolates sampled from Nanchang in nearby Jiangxi Province. The 1996 hemagglutinin and neuraminidase appeared closest to those of H5N3 and H1N1 isolates from Japan. In the months before the outbreak in Hong Kong several of the proteins were again replaced by way of recombination, this time via strains of H9N2 and H6N1 (Guan et al., 1999; Hoffmann et al., 2000). H5N1 outbreaks in the years that followed Hong Kong emerged by still more recombination (Li et al., 2004). The sociogeographic mechanisms by which the various segments first converged (and were repeatedly shuffled) in Guangdong remain to be better outlined. The results so far do indicate the spatial expanse over which reassortants originate may be greater than Kennedy Shortridge, or anyone else, previously outlined. But genomic origins tell us little how this particular complement led to a virus that *locally evolved* such virulence other than showing the genetic variation upon which the virus can draw.

A closer look at Guangdong's drastically shifting socioeconomic circumstances, then, appears necessary in better illuminating the local conditions that selected for such deadly pathogens so easily spread; not only H5N1, but a diverse viral portfolio, including influenza A (H9N2) (Liu et al., 2003), H6N1 (Cheung et al., 2007), and SARS (Poon et al., 2004). What exactly are the 'crucial parameters' for the area's disease ecosystem? What are the mechanisms by which changes in southern China's human-animal composite lead to regular viral pulses emanating out to the rest of China and the world? Why Guangdong? Why 1997 and thereafter?

6.4.5 700 million chickens

We begin with the death of Mao and the rehabilitation of Deng Xiaoping. In the late 1970s, China began to move away from a Cultural Revolution policy of self-sufficiency, in which each province was expected to produce most foods and goods for its own use. In its place, the central government began an experiment centered about a reengagement with international trade in Special Economic Zones set up in parts of Guangdong (near Hong Kong) and Fujian (across from Taiwan), and later the whole of Hainan Province. In 1984, 14 coastal cities-including Guangzhou and Zhanjiang in Guangdong-were opened up as well although not to the extent of the economic zones (Tseng and Zebregs, 2003).

By macroeconomic indicators favored by establishment economists, the policy was a success. Between 1978 and 1993 China's trade-to-GNP ratio grew from 9.7% to 38.2% (Perkins, 1997). Most of this growth stemmed from manufactured goods produced by foreign-funded joint ventures and township and village enterprises (TVE) allowed greater autonomy from central control. Foreign direct investment (FDI) increased from nothing to US $45 billion by the late 1990s, with China the second greatest recipient after the US. Sixty percent of the FDI was directed to cheap-labor Chinese manufacturing. Given the extent of China's small-holder farming, little FDI was initially directed to agriculture (Rozelle et al., 1999).

That has since begun to change. Through the 1990s poultry production grew at a remarkable 7% per year (Hertel et al., 1999). Processed poultry exports grew from US $6 million in 1992 to US$774 million by 1996 (Carter and Li, 1999). The Interim Provisions on Guiding Foreign Investment Direction, revised in 1997, aim to encourage FDI across a greater expanse of China and in specific industries, agriculture included (Tseng and Zebregs, 2003). China's latest 5-year plan sets sights on modernizing agriculture nationwide (Tan and Knor, 2006). Since China joined the WTO in 2002, with greater obligations to liberalize trade and investment, agricultural FDI has doubled (Whalley and Xin, 2006). But much opportunity for AgFDI remains available to a wider array of sources of investment. By the late 1990s, Hong Kong and Taiwan's contribution to China's FDI had declined to 50% of the total, marking an influx of new European, Japanese, and American investment.

In something of a bellwether, in August, 2008, days before the Olympics, U.S. private equity investment firm Goldman Sachs bought ten poultry farms in Hunan and Fujian for US$300 million (Yeung, 2008). Although the image of a band of New York brokers knee-deep in chicken shit may prompt a cackle, Goldman Sachs has contracted third parties to run the farms. The outright ownership appears a step beyond the joint ventures in which Goldman Sachs had until then participated. Goldman Sachs already holds a minority stake in Hong Kong-listed China Yurun Food Group, a mainland meat products manufacturer, and 60% of Shanghai-listed Shuanghui Investment and Development, another meat packer. Goldman Sachs' new purchase, further up the filiere, signals a shift in the global fiscal environment. The firm has voted with its feet, deftly moving out of high-risk US mortgages and, during a global food crisis, into Chinese farming.

Guangdong remains at the cutting edge of the economic shift. It hosted the central government's first efforts at internationalizing the rural economy (Zweig, 1991; Johnson, 1992; Xueqiang et al., 1995). Starting in 1978, agricultural production was redirected from domestic grain to Hong Kong's market. Hong Kong businesses invested in equipment in return for new output in vegetables, fruit, fish, flowers, poultry and pig. In something of a return to its historical role, Hong Kong ('the front of the store') also offered Guangdong ('the back of the store') marketing services and access to the international market (Sit, 2004; Heartfield, 2005). In a few short years Guangdong's economy again became entwined with and dependent upon Hong Kong's economic fortunes. And vice versa. As of the Hong Kong outbreak, investment in China comprised 4/5 of Hong Kong's FDI outflow (Heartfield, 2005). Much of Hong Kong-funded production is now conducted in Guangdong, with Hong Kong's industrial base increasingly hollowed out as a result.

Eighty-five percent of the agricultural FDI brought in during the 1990s was funneled into Guangdong and several of the other coastal provinces (Rozelle et al., 1999). Guangdong was allowed to invest more in its transportation infrastructure, an invitation for further investment. Many of the province's companies were allowed to claim 100% duty drawbacks. Guangdong also developed trading arrangements with many of the 51 million Chinese overseas (Gu et al. 2001, Heartfield 2005). As a class the expatriates, nearly 200 years abroad, control large percentages of regional market capital, including in Indonesia, Thailand, Vietnam, the Philippines, Malaysia, and Singapore. At the time of the first H5N1 outbreaks overseas Chinese collectively comprised the group with the greatest investment in mainland China (Haley et al. 1998).

As a result of the area-specific liberalization, Guangdong accounted for 42% of China's total 1997 exports and generated China's largest provincial GDP (Lin, 2000; Gu et al., 2001). Of the coastal provinces, Guangdong hosted the greatest concentration of joint-venture export-oriented firms, with the lowest domestic costs for each net dollar of export income (Perkins 1997). Guangdong's three free economic zones (Shenzhen, Shantou and Zhuhai) boasted an export-to-GDP ratio of 67%, compared to a national average of 17%.

By 1997, and the first H5N1 outbreak in Hong Kong, Guangdong, home to 700 million chickens, served as one of China's top three provinces in poultry production (Organisation for Economic Co-operation and Development, 1998). Fourteen percent of China's farms with 10,000 or more broilers were located in Guangdong (Simpson et al., 1999). Guangdong's poultry operations were by this point technically modernized for breeding, raising, slaughtering, and processing birds, and vertically integrated with feed mills and processing plants. AgFDI helped import grandparent genetic stock, support domestic breeding, and introduce superior nutrition feed milling/mixing (Rozelle et al., 1999). Production has been somewhat constrained by access to interprovincial grain and the domestic market's preference for native poultry breeds less efficient at converting feed. Of obvious relevance, production also suffered from inadequate animal health practices.

The rate and magnitude of poultry intensification poultry appears to have combined with the pressures placed on Guangdong wetlands by industry and a burgeoning human population to squeeze a diversifying array of influenza serotypes circulating year-round through something of a virulence filter. The resulting viral crop-for 1997, H5N1 by molecular happenstance-is exported out by easy access to international trade facilitated in part by expatriate companies.

Guangdong's ascension wasn't without its detractors. Domestic producers in Hong Kong competed with Hong Kong-Guangdong joint ventures for export licenses (Zweig, 1991). Landlocked provinces meanwhile chafed at the liberalization the central government proffered coastal provinces alone. With so much domestic currency on hand, the coastal provinces could outcompete inland provinces for livestock and grain produced by the inland's own TVEs. The coastal provinces were able to cycle their competitive advantage by turning cheap grain into more profitable poultry or flat-out re-exporting the inland goods, accumulating still greater financial reserves. At one point rivalries became so intense that Hunan and Guangxi imposed trade barriers upon interprovincial trade. The central government's efforts to negotiate interprovincial rivalries included spreading liberalization inland (Tan and Khor, 2006). Provinces other than Guangdong and Fujian began to become entrained into market agriculture, albeit at a magnitude still outpaced by their coastal counterparts. Industrial poultry's expanding extent increases the geographic scope for H5N1's emergence and may explain the roles Yunnan and Hunan appear to have played in serving up H5N1 abroad.

An additional source of conflict, often forgotten in the cacophony of macroeconomic indicators, requires comment-the Chinese people themselves. China's state capitalism has induced such a polarization of wealth that, along with threatening its own economic growth, impoverishes hundreds of millions of Chinese. In engaging in internally imposed structural adjustment China has largely turned away from its real and ideological investment in the health and wellbeing of its population (Hart-Landsberg and Burkett, 2005a). Tens of millions of state industrial workers have been laid off. Labor income as a share of Chinese GDP fell from about 50% in the 1980s to under 40% by 2000 (Li, 2008). FDI and private companies-under no obligation to offer housing, healthcare, or retirement benefits-are used to discipline Chinese workers who were long used to a living wage, basic benefits, and job protections (Hart-Landsberg and Burkett, 2005b). Discipline, however, does not always take. Protests running now into the tens of thousands, some turning into riots requiring army deployment, have battered provincial governments accused of corruption, land confiscation, expropriating state assets, wage theft, and pollution. In something of an ironic twist, in defending foreign capital against its own people China's communist leadership has taken on the role of the comprador class it defeated in 1949 (Heartfield, 2005).

Farmers have been particularly hard hit by the government's capitalist turn. While decollectivization of agricultural land to household control propped up by governmental price supports led to a doubling in rural incomes by 1984, rural infrastructure and attendant social support deteriorated (Hart-Landsberg and Burkett, 2005a). In the late 80s, agricultural incomes stagnated, eaten away by inflation and a decline in price supports. Families began to abandon farming for informal industrial work in the cities. There, many rural migrants are treated as a reviled caste, discrimination codified by levels of officially designated migrant status and with attendant effects on income (Fan 2001). China's macroeconomic growth has been unable to absorb many of the 100 million migrants.

Urbanization meanwhile has diffused out to the rural regions, eating up peasant land. One million Chinese hectares have been converted from agriculture to urban use (Davis, 2006). Remote sensing shows from 1990 to 1996 13% of agricultural land in a ten-county region in Guangdong's Pearl River delta was converted into non-agricultural use, in all likelihood China's most rapid conversion (Seto et al., 2000). Rural towns have been transformed into growing industrial cities, some supporting populations tipping a million people (Lin, 1997).

The termination of the commune system has left hundreds of millions of peasants without access to medical care and health insurance (Shi 1993). Universal health coverage has degraded to 21% of the rural population insured (French, 2006). The number of affordable doctors has precipitously declined. Infant mortality has risen across many provinces. Rural public health has largely collapsed. Hepatitis and TB are now widespread. HIV incidence has increased in several southeastern provinces, Guangdong included (Tucker et al., 2005). STI incidence by province is correlated with immigration associated with surplus men from rural regions separated from their families. Multitudes of malnourished and immunologically stressed peasants cycle-migrating back and forth from what may be the geographic origins of an influenza pandemic would appear to compromise World Health Organization plans for intervening at any new infection's source.

6.4.6 Asian financial flu

It is hard to talk of 1997 without mentioning two events of geopolitical significance. On July 1 Hong Kong, long a British colony, was officially transferred to China as a Special Administrative Region, the first in a series of steps to full integration to be undertaken up through 2047. The next day the Bank of Thailand floated the baht off the US dollar. The baht had been hammered by currency speculation and a crippling foreign debt. International finance fled the baht and soon, with the economic strength of Thailand's neighbors also under suspicion, from other regional currencies. The FDI-dependent economies of the Philippines, Malaysia, Indonesia, Taiwan, and South Korea suffered in the ensuing wave of devaluation. The rest of the world too felt the effects of the infectious 'Asian flu,' as the crisis came to be called, with stock markets worldwide free-falling in response. Although the transfer of Hong Kong to China and the Asian financial crisis followed the first outbreaks of avian influenza in March, the events marked long-brewing shifts in regional political economy with apparent impact on viral evolution and spread.

Hong Kong's role in China's internally imposed structural adjustment, as we explored above, is amply documented. The intensification of Guangdong poultry went hand in hand with the ongoing transformation of the province's border with Hong Kong (Breitung 2002). The resulting poultry traffic, however, is in no way unidirectional. Hong Kong exports to mainland China large amounts of poultry, fruits, vegetables, nuts, oilseeds, and cotton (Carter and Li 1999). There is too a large illegal trade. At the time of the outbreak, Hong Kong chicken parts smuggled into China alone may have amounted to over US$300 million per year (USTR, 1998; Carter and Li, 1999). Hong Kong is clearly less a victim of Guangdong's avian influenza ecology, as often portrayed, than a willing participant.

Meanwhile, the financial crisis slowed China's economy. But because of the central government's intervention China avoided the worst of the flu (Lin, 2000). By staking billions in public works and loans, China kept the economic

engine primed in the face of slowing exports. Prophetically, four years previous, the central government introduced fiscal austerity measures to cool off
inflation and the possibility of an overheated economy. An associated regulation package was initiated to control the kind of short-term speculation that
would soon strain China's regional neighbors. The central state maintains
tight control over the macroeconomy, capital flows, and corporate structure
even as it cedes much of the day-to-day operations to provincial authorities.
Concomitantly, China's economy is more than export-driven. Even as austerity leaves millions of Chinese destitute in its wake (Hart-Landsberg and
Burkett, 2005), the domestic economy continues to grow, albeit increasingly
dependent on luxury goods and real estate speculation. Finally, exports out
of China were until the crisis largely destined for East and Southeast Asia.
During the crisis' aftermath China redirected more of its trade to Europe,
North America, Africa, Latin America, and Oceania. China, then, was able
to maintain a trade surplus, retain foreign investment, and prop the yuan
against the fiscal buffet from abroad.

At the same time, China was something more than a bystander to the
crisis. Its economy's growing size and hemispheric reach may have exposed its
neighbors to the worst excesses of the neoliberal model (Hart-Landsberg and
Burkett, 2005a; Tan and Khor, 2006). In attracting FDI at rates above and
beyond those of its neighbors, China has become the prime exporter in the
region: textiles, apparel, household goods, televisions, desktop computers, an
increasing array of high-end electronics-you name it. The smaller economies
are forced to restructure production in such a way as to complement China's
increasingly diverse commodity output, in a type of regional division-of-labor.
China's transnational impact on supply lines forces each country to depend
on producing a smaller array of parts to be put together in China for final
export.

The resulting economies are more dependent on what few foreign multinationals they are able to attract. The company town becomes the company
country. Such economies are more 'brittle'-less robust in reacting to and re-
orienting around downturns in any single industry, a particularly pernicious
problem as the US begins to falter in its role of importer of last resort. The
capital flight exposes countries to the temptations of currency speculation. To
attract additional investment, establishment economists declare these countries, once burned by such speculation, must now remove remaining barriers
to the movement of money, goods, and capital, leaving domestic production
unprotected, the very conditions that brought about the 1997 crisis in the
first place.

It would appear bird flu and the financial flu are intimately connected,
their relationship extending beyond serendipitous analogy. Although agriculture has until recently been less export-dependent than manufacturing, in part
from its perishability and now endangered trade protections (Hertel et al.,
2000), there are already a number of epidemiological ramifications. These include a geographically expanding and intensifying poultry production, greater

exposure to transnational poultry, wider illegal poultry trade, and a truncation in animal health infrastructure by austerity measures domestically imposed in return for international loans (Rweyemamu et al., 2000). More acutely, the aftermath of the financial flu may have also provided China a window for expanding regional poultry exports. A hypothesis worth testing is that some of these shipments seeded avian influenza outbreaks abroad.

How do we operationalize this model? How do we determine whether transnational companies breed and spread avian influenza? Identifying poultry crates carrying H5N1 country-to-country remains a difficult, but important, task (Kilpatrick et al., 2006). Tracing pathogens through commodity chains is increasingly an important topic of study and mode of intervention (Duffy et al., 2008). One difficulty centers about the willingness of government regulators to inspect poultry plants, including conditions under which pathogen virulence may evolve. At the same time, there is a danger such efforts, once successful, may detract from the larger political ecology that shapes avian influenza evolution. With billions annually at stake, a few unlucky contract farmers or truck drivers may be sacrificed to protect a system stretching across a hemisphere's interlocking markets. We've explored here the possibility a deadly avian influenza is an unintended but not unexpected accessory to multinational efforts to export a growing portfolio of Chinese agricultural commodities. The problem of avian influenza is more than a police matter. It is systemic, buried deep in political tissue.

6.4.7 Layers of complication

Ending poultry production as we know it could make a great difference in Guangdong as elsewhere. But there are additional layers of complication. There is no easy one-to-one relationship between poultry density and H5N1 outbreak at a variety of spatial scales. Across Asia, some areas where outbreaks have occurred support comparatively few poultry, while other areas with millions of chickens have been so far left untouched. There is something of a stochastic component to disease spread. Epidemics start somewhere, in this case in southern China, and take time to wend their way elsewhere, starting with regions nearby and, in part by due cause and in part by chance, farther abroad. There are, however, demonstrable causes other than those inside the poultry industry.

Thailand offers one such example. As mapped by ecologist Marius Gilbert and colleagues (Gilbert et al., 2006; Gilbert et al., 2008), the distribution of Thai broilers and backyard poultry appear little associated with H5N1 outbreaks. Local outbreaks appear better fitted to the densities of ducks that are allowed to graze freely outside. After harvests these ducks are brought in to feed on the rice that is left over on the ground. Satellite pictures show rice harvests matching duck densities. The more annual rice crops, the more ducks (and the greater the association with H5N1 outbreaks). It seems these ducks, free to graze outdoors, exposed to migratory birds, and tolerant of a wider range of influenzas, serve as epidemiological conduits for infecting nearby poultry. While a rather ingenious agricultural practice, raising a cohort of ducks on fallen waste rice may carry serious epidemiological overhead. Double- and even triple-cropping is practiced in other avian influenza zones, including southeastern China, the final stretches of the Xun Xi River, the Ganges floodplain, and on the island of Java (Leff et al., 2004).

We have, then, an integrated viral ecology with highly complex dependencies. The variety of farming practices, for one, splits a-twain a number of facile dichotomies. There is a panoply of farm types, beyond the rough polarities of 'small' and 'large.' In Thailand alone there are closed-off farms, open structures with netting to block passerine birds, the aforementioned free-grazing ducks, and backyard poultry (Songserm et al., 2006).

Even then, such a taxonomy implies a compartmentalization often absent in the field. On a recent trip to Lake Poyang in Jiangxi Province, China, a team of international experts discovered an astonishing farming ecology in which domesticated free-range ducks fed in fields, bathed in local estuaries, swam in the lake, and intermingled and presumably interbred with wild waterfowl. Some flocks daily commuted across dikes from their sheds to the open water and back. The epidemiological implications are obvious. Indeed, the facility by which pathogens spread and evolve in the area is of an order that, according to local farmers, chickens cannot be raised around the lake. For some poultry species the region is epidemiologically radioactive.

Absent too from the taxonomy are profound structural changes imposed by economic pressures upon world farming (Weis, 2007). For the past three decades, the International Monetary Fund and the World Bank have made loans to poorer countries conditioned on removing supports for domestic food markets. Small farmers cannot compete with cheaper corporate imports subsidized by the Global North. Many farmers either give up for a life on periurban margins or are forced to contract out their services-their land, their labor-to livestock multinationals now free to move in (Manning and Baines, 2004; Lewontin, 2007). The World Trade Organization's Trade-Related Investment Measures permit foreign companies, aiming to reduce production costs, to purchase and consolidate small producers in poorer countries (McMichael, 2006). Under contract, small farmers must purchase transnational-approved supplies and are given no guarantee their birds will be bought back by their transnational partner. The new arrangements belie the superficial distinction that has been made between factory farms exercising 'biosecurity' on the one hand and small farmers whose flocks are exposed to the epidemiological elements. Factory farms ship day-old chicks to be raised piecework by small farmers. Once grown (and exposed to migratory birds), the grown birds are shipped back to the factory for processing. The violation of 'biosecurity' appears built directly into the industrial model.

A third complication is the historical shift in the relationship between nature and farming. Maps in Phongpaichit and Baker (1995) show since 1840 Thailand has been transformed from primary wilderness into an agricultural state, a veritable bread-basket. Agriculture's new girth comes at the expense of wetlands worldwide, either out-and-out destroyed, polluted, or irrigated dry. The latter abuse serves as another basis for conflicts between agribusiness and small farmers. Socially stratified power struggles over the Chao Phraya basin have wracked Thailand for hundreds of years (Molle, 2007).

Wetlands have traditionally served as *Anatidae* migration pit stops (Lemly et al., 2000). A growing literature shows many migratory birds are no sitting ducks and have responded to the destruction of their natural habitat. Geese, for example, display an alarming behavioral plasticity, adopting entirely new migratory patterns and nesting in new types of wintering grounds, moving from deteriorating wetlands to food-filled farms. The shift has for some populations substantially increased their numbers (Jeffries et al. 2004, Van Eerden et al., 2005). The population explosions have initiated a destructive feedback in which the swarms of farm-fed migratory birds overgraze their Arctic breeding grounds to the point the tundra is transformed into a mud pit. In the course of colonizing our planet's natural habitats-some 40% of the world's usable land now supports agriculture-we may have unintentionally expanded the interface between migratory birds and domestic poultry.

Clearly agribusiness, structural adjustment, environmental destruction, climate change, and the emergence of avian influenza are more tightly integrated than previously thought.

6.4.8 The political will for an epidemiological way?

Guangdong may only represent the front of a socioecological transformation spreading across much of southern China, as well as across much of the populated world. The origins of highly pathogenic H5N1 are multifactorial, with many countries and industries and sources of environmental damage at fault. Can we then place blame on the country, say, Indonesia or Vietnam or Nigeria, from which a human-to-human pandemic might first emerge? Should we blame China for repeatedly seeding outbreaks regionally and internationally? Should we blame Hong Kong for offshore farming? Or should we blame the United States, where the industrial model of vertically integrated poultry first originated, with thousands of birds packed in as so much food for flu? The answers are yes, yes, yes, and yes. Blame, much as the problem itself, must be distributed about its multiple levels of social and ecological organization.

To break avian influenza's back, or at the very least promote some sort of sustainable epidemiological mitigation, a number of radically invasive changes are required, changes that challenge core premises of present political economy, neoliberal and state capitalist alike. Whether there exists the political will to change is an open question. Denial, jockeying, and obfuscation are presently rampant. Chinese officials have expended much effort in flat-out denying responsibility for avian influenza (Wallace, 2007) or, in the epidemiological equivalent of the American practice of paying off the families of collateral damage without admitting guilt, offering small sums to affected countries.

In 2007, China donated US $500,000 to Nigeria's effort to fight avian influenza. Never mind that Nigeria would never have needed the aid if China hadn't infected it with avian influenza in the first place. The Qinghai-like strain Nigeria now hosts first originated in southern China. Meanwhile, the US and EU, laying undue blame on a stubborn Indonesia unwilling to share H5N1 samples, have blocked efforts to reform a system of worldwide vaccine production that rewards pharmaceutical companies and the richest populations at the expense of the poorest (Hammond, 2007, 2008).

What must be done to stop avian influenza, if the political will is found by, or forced upon, governments worldwide? In the short term, small farmers must be fairly compensated for poultry culled in an effort to control outbreaks. Poultry trade must be better regulated at international borders (Kilpatrick et al., 2006; Wallace and Fitch, 2008). The world's poor must be provided epidemiological assistance, as well as vaccine and antiviral at no cost (Cristalli and Capua, 2007; Ferguson 2007). Structural adjustment programs degrading animal health infrastructure in the poorest countries must be terminated.

For the long term, we must end the poultry industry as we know it. Avian influenza now emerges by way of a globalized network of corporate poultry production and trade, wherever specific strains first evolve. With poultry batches whisked from region to region-transforming spatial distance into just-in-time expediency (Harvey, 1982/2006)-multiple strains of avian influenza are continually introduced into localities filled with populations of susceptible birds. Such domino exposure serves as the fuel for the evolution of viral virulence. In overlapping each other along the links of agribusiness's transnational supply chains, strains of avian influenza also increase the likelihood they can exchange genomic segments to produce a recombinant of pandemic potential. In addition to the petroleum wasted and the loss of local food sovereignty there are epidemiological costs to the geometric increase in food miles.

We must instead devolve much of the production to regulated networks of locally owned farms. While the argument has been made that corporate chicken supplies the cheap protein many of the poorest need, the millions of small farmers who feed themselves (and many millions more) would never have needed such a supply if they hadn't been pushed off their lands in the first place. A reversal need not be solely an anachronistic turn to the small family farm, but can include domestically protected farming at multiple scales (Levins, 1993, 2007; Brown and Getz, 2008). Farm ownership, infrastructure, working conditions, and animal health are inextricably linked. Once workers have a stake in both input and output-the latter by outright ownership, profit sharing, or the food itself-production can be structured in such a way that respects human welfare, and, as a consequence, animal health. With local farming, genetic monocultures of domesticated bird which promote the evolution of virulence can be diversified back into heirloom varieties that serve as immunological firebreaks. The economic losses bird flu imposes upon global poultry can be tempered: fewer interruptions, eradication campaigns, price jolts, emergency vaccinations, and wholesale flock repopulations (Van Asseldonk et al., 2006). Rather than jury-rigged with each outbreak, restrictions on bird movement are built naturally into the independent farm model.

The devil of such a domain shift is in its details. Richard Levins (2007), with decades experience collaborating with local researchers and practitioners on ecological approaches to Cuban agriculture and public health, summarizes some of the many adjustments a new agriculture may require,

> "Instead of having to decide between large-scale industrial type production and a "small is beautiful" approach a priori, we saw the scale of agriculture as dependent on natural and social conditions, with the units of planning embracing many units of production. Different scales of farming would be adjusted to the watershed, climatic zones and topography, population density, distribution of available resources, and the mobility of pests and their enemies.
>
> The random patchwork of peasant agriculture, constrained by land tenure, and the harsh destructive landscapes of industrial farming would both be replaced by a planned mosaic of land uses in which each patch contributes its own products but also assists the production of other patches: forests give lumber, fuel, fruit, nuts, and honey but also regulate the flow of water, modulate the climate to a distance about ten times the height of the trees, create a special microclimate downwind from the edge, offer shade for livestock and the workers, and provide a home to the natural enemies of pests and the pollinators of crops. There would no longer be specialized farms producing only one thing. Mixed enterprises would allow for recycling, a more diverse diet for the farmers, and a hedge against climatic surprises. It would have a more uniform demand for labor throughout the year."

The scale and practice of agriculture must be flexibly integrated into the region's physical, social and epidemiological landscapes. At the same time, it need be acknowledged that under such an arrangement not all parcels will be routinely profitable. Whatever reductions in income farms accrue in protecting the rest of the region must be offset by regular redistributive mechanisms (Richard Levins, personal communication).

Transforming the business of farming so broadly is likely only one of many steps necessary to stop bird flu and other pathogens. For one, migratory birds, which serve as a fount of influenza strains, must concomitantly be weaned off agricultural land where they cross-infect poultry. To do so, wetlands worldwide, waterfowl's natural habitat, must be restored. Global public health capacity must also be rebuilt (Garrett, 2001). That capacity is only the most immediate bandage for the poverty, malnutrition, and other manifestations of structural violence that promote the emergence and mortality of infectious diseases, including influenza (Kim et al., 2000; Farmer, 2004). Pandemic and inter-pandemic flu have the greatest impact on the poorest (Davis, 2005). As for many pathogens, particularly for such a contagious virus, a threat to one is a threat to all.

Only once these objectives are fulfilled will we be able to better cover ourselves against H5N1 and the other influenza serotypes now lining up like hurricanes brewing in the Atlantic.

7

Final Remarks

The previous chapter suggests that evolutionary transformations in human pathogens can be driven by ecosystem resilience shifts constituting reductionist interventions or changes in predatory socioeconomic structures. Reductionist interventions are, for the most part, themselves the product of particular socioeconomic perspectives: Use cheap 'magic bullets' to avoid the necessity of comprehensive social change. But ecology drives evolution, and these approaches are guaranteed to farm infectious disease, producing ever more efficient microbial and viral predators on human populations. Although a contrary strategy is clearly possible – social and economic reforms designed to lessen pathogen virulence and spread – reductionist interventions presently serve primarily as a political firebreak against effective public health programs: Tamiflu and body bags, for instance, insulate the international poultry industry.

The consequences will likely far transcend the development of pathogen drug resistance, and may drive life history strategies for organisms which operate at multiple scales. Reductionist interventions can select for holistic diseases far beyond our coping strategies. Rapacious economic practice can create new, or enlarge old, niches for pathogens, often through spatial or economic dislocation.

The alternative strategy is one of 'Integrated Pathogen Management', essentially an ecologically informed return to the basic principles of public health, to public policies and economic practices that synergistically interrupt transmission of human pathogens through coordinated interventions targeted at different scales of space, time, and population. The scientific and administrative tools needed for such a program have long been with us, but the political ability to use them remains thwarted by powerful economic interests, and by historical trajectories of marginalization, inequality, and deadly conflict that so often become useful tools for those interests.

Wallace and Fullilove (2008) have shown in some detail how American Apartheid interacted with HIV's 'basic biology' to undercut attempts to control and contain AIDS in the United States. Here we have indicated something

R. Wallace et al., *Farming Human Pathogens*, DOI 10.1007/978-0-387-92213-3_7,
© Springer Science+Business Media, LLC 2009

of the cost of that failure in terms of the 'farming' of multiple drug resistant HIV. Avian influenza presents another, perhaps more deadly, example of the large scale cultivation of a human pathogen.

This is not, in all, a particularly new perspective. Ecosystem methods are widely touted as tools for controlling emerging infections, (e.g., Garrett and Cox, 2008). Our contribution is to apply recently developed formal theoretical tools linking evolutionary transformations to ecological resilience domain shifts – the results of Wallace and Wallace (2008). Ecosystem transformations entrain the evolution of human pathogens, just as they do other forms of evolutionary change. From a theoretical perspective, the basic point is the inevitability of mesoscale-driven punctuation in generalized coevolutionary interactions. Thus evolution, ecosystem resilience, and cognitive phenomena, which can all be (at least crudely) represented by information sources, are inherently subject to punctuated equilibrium dynamics. This can involve each individually, as well as their interactions.

Holling (1992) finds that ecosystems are, by and large, controlled and organized by a small number of key plant, animal, and abiotic processes that structure the landscape at different scales. He invokes an entrainment hypothesis, that within any one ecosystem, the periodicities and architectural attributes of the critical structuring processes will establish a nested set of periodicities and spatial features that become attractors for other variables. He argues that the degree to which small, fast events influence larger, slower ones is critically dependent upon mesoscale disturbance processes.

Our lowest-common-denominator information-theoretic approach to coevolutionary interaction between genes, embedding ecosystem, and cognitive process identifies ecosystem phenomena as the driving mesoscale: cognitive phenomena are much faster, and (for large animals) genetic change much slower. The mesoscale resonance argument of section 2.12 provides a formal basis for these assertions. That is, punctuated changes in ecosystem structure, the traditional purview of ecological resilience, appear able to entrain both Darwinian genetic and cognitive phenomena – including gene expression – triggering similarly punctuated outcomes, on top of the punctuation naturally inherent to these information systems.

Thus, while discontinuous phase transitions are 'natural' at all scales of biological information process, we argue that punctuated changes in an embedding ecosystem resilience regime will be particularly effective at entraining faster cognitive and slower Darwinian genetic structural phenomena. In particular, punctuated changes in ecosystem structure can write images of themselves onto genetic sequence structure in a punctuated manner, resulting in punctuated population extinction and/or speciation events on geologic timescales, and in sudden changes in gene expression and other cognitive phenomena on more rapid timescales, including the evolution of human pathogens. The case histories of chapter 6 suggest that for broad classes of human pathogens these effects may be quite rapid, with public policy, reductionist interventions, socioeconomic structure and process, and their many synergisms, playing the key mesoscale roles.

This is, again, not an entirely new approach. Laland et al. (1999) have used a different methodology to reach similar conclusions. In their view there is increasing recognition that all organisms modify their environments through a process they characterize as 'niche construction'. Such modifications can have profound effects on the distribution and abundance of organisms, the influence of keystone species, the control of energy and material flows, residence and return times, ecosystem resilience, and specific trophic relationships. The consequences of environment modification by organisms, however, are not restricted to ecology, and organisms can affect both their own and each other's evolution by modifying sources of natural selection in their environments. They cite Lewontin's work, which points out that many of the activities of organisms, such as migration, hoarding of food resources, habitat selection, or thermoregulatory behavior, are adaptive precisely because they dampen statistical variation in the availability of environmental resources.

Laland et al. (1999) argue that, hitherto, it has not been possible to apply evolutionary theory to ecosystems, because of the presence of nonevolving abiota in ecosystems. They suspect this obstacle has been largely responsible for preventing the full integration of ecosystem ecology with population-community ecology. However, in their view, adding the new process of niche construction to the established process of natural selection enables the incorporation of both abiotic environmental components and interactions among populations and abiota in ecosystems into evolutionary models an approach equally applicable to both population-community ecology and ecosystem-level ecology.

Odling-Smee et al. (1995, 2003) have discussed these matters from the perspective of Lewontin, who has argued that the 'metaphor of adaptation' should be replaced by a 'metaphor of construction'. However, the acceptance of Lewontin's position, they state, demands more than just semantic adjustments to evolutionary theory. Niche construction changes the dynamic of the evolutionary process in fundamental ways because it precludes a description of evolutionary change relative only to autonomous environments. Instead, evolution now consists of endless cycles of natural selection and niche construction. Equally, it is no longer tenable from their perspective to assume that the only way organisms can contribute to evolutionary descent is by passing on fit or unfit genes to their descendants relative to their environments, because they can also pass on modifications in those environments that are better or worse suited to their genes. Adaptation becomes a two-way street in this theory.

More recently, Dercole et al. (2006) have addressed the problem using their version of equation (4.8) to produce very complex dynamical patterns, focusing on eco-evolutionary dynamics in communities containing 'slow' and 'fast' populations, which allows relaxing assumptions of ecological equilibrium.

Whitham et al. (2006), in parallel with our approach, take a genetic framework associated with ecologically-dominant keystone species to examine what they call community and ecosystem phenotypes. They ask whether heritable traits in a single species can affect an entire ecosystem. Recent studies, they claim, show that such traits have predictable effects on community structure and ecosystem processes. Because these community and ecosystem phenotypes have a genetic basis and are heritable, they claim it is possible to apply the principles of population and quantitative genetics to place the study of complex communities and ecosystems within an evolutionary framework. This could, they assert, allow us to understand, for the first time, the genetic basis of ecosystem processes, and the effect of such phenomena as climate change and introduced transgenetic organisms on entire communities.

Whitham et al. (2006) go on to define community evolution as a genetically based change in the ecological interactions that occur between species over time.

Here, by contrast, although we too focus on keystone scales, our particular innovation has been to reduce the dynamics of genetic inheritance, ecosystem persistence, and gene expression to a least common denominator as information sources operating at markedly different rates, but coupled by crosstalk into a broadly coevolutionary phenomenon marked at all scales by emergent 'phase transition' phenomena generating patterns of punctuated equilibrium.

Invocation of equivalence class arguments leads naturally into groupoid structures and related topological generalizations, including Morse theory. Taking a 'mean number' rather than the mean field approach generates a qualitatively different class of exactly solvable models, based on giant component phase transitions in networks. Hybrids of the two are possible, and evolutionary process is unlikely to be at all constrained by formal mathematical tractability.

Higher cognitive phenomena – 'farming' by embedding cultural structures – has now become the most powerful determinant of human pathogen evolution.

We conclude with E.C. Pielou's (1977) important warning regarding the kind of ecological modeling we do here:

"...[Mathematical models] are easy to devise; even though the assumptions of which they are constructed may be hard to justify, the magic phrase 'let us assume that...' overrides objections temporarily. One is then confronted with a much harder task: How is such a model to be tested? The correspondence between a model's predictions and observed events is sometimes gratifyingly close but this cannot be taken to imply the model's simplifying assumptions are reasonable in the sense that neglected complications are indeed negligible in their effects...

In my opinion the usefulness of models is great... [however] it consists *not in answer questions but in raising them*. Models can be used to inspire new field investigations and these are the only source of new knowledge as opposed to new speculation."

The principal model-based speculation of this work is that, via the mechanisms of section 2.12, mesoscale ecosystem resilience shifts can entrain punctuated events of gene expression and other cognitive phenomena on more rapid time scales, and slower genetic selection-induced changes, triggering punctuated equilibrium Darwinian evolutionary transitions on broader time scales. For microbial or viral pathogens, chapter 6 argues that this process may be quite rapid indeed.

The model we have invoked, unlike most related work, is a statistical one in which the asymptotic limit theorems of information theory impose necessary conditions on the behavior of ecosystems, Darwinian genetic selection, and gene expression. These necessary conditions, as the Central Limit Theorem does for regression theory, permit the construction of empirical models which can be fitted to data. Scientific inference is not in the model fitting itself, but rather, in the comparison of similar systems under different conditions, and the comparison of different systems under similar conditions. This semi-empirical approach is, perhaps, what most differentiates our developments from other attempts to model biological processes. We can, at best, impose necessary conditions through our formal development. The real science must then be done by real experiment.

For human populations in particular, several other layers of information sources, those of Lamarckian culture, and of individual and group consciousness and learning, become manifest, producing a rich stew of complicated phenomena (Wallace, 2004, 2005b; Wallace and Fullilove, 2008).

The coevolutionary interaction between our cultural structures and our pathogen predators plays out these dynamics in real time. Absent widespread, comprehensive, progressive social reform, the results are likely to be globally catastrophic, in no way restricted to the disenfranchised poor. The rapacity of the pharmaceutical, real estate, agribusiness, and other international elites – the global nomenklatura – will write an image of itself across human populations as a relentless parade of plagues.

8

Mathematical Appendix I

8.1 The Shannon-McMillan Theorem

According to the structure of the underlying language of which a message is a particular expression, some messages are more 'meaningful' than others, that is, are in accord with the grammar and syntax of the language. The Shannon-McMillan or Asymptotic Equipartition Theorem, describes how messages themselves are to be classified.

Suppose a long sequence of symbols is chosen, using the output of the random variable X above, so that an output sequence of length n, with the form

$$x_n = (\alpha_0, \alpha_1, ..., \alpha_{n-1})$$

has joint and conditional probabilities

$$P(X_0 = \alpha_0, X_1 = \alpha_1, ..., X_{n-1} = \alpha_{n-1})$$

$$P(X_n = \alpha_n | X_0 = \alpha_0, ..., X_{n-1} = \alpha_{n-1}).$$

Using these probabilities we may calculate the conditional uncertainty

$$H(X_n | X_0, X_1, ..., X_{n-1}).$$

The uncertainty of the *information source*, $H[\mathbf{X}]$, is defined as

$$H[\mathbf{X}] \equiv \lim_{n \to \infty} H(X_n | X_0, X_1, ..., X_{n-1}).$$

(8.1)

In general

$$H(X_n|X_0, X_1, ..., X_{n-1}) \leq H(X_n).$$

Only if the random variables X_j are all stochastically independent does equality hold. If there is a maximum n such that, for all $m > 0$

$$H(X_{n+m}|X_0, ..., X_{n+m-1}) = H(X_n|X_0, ..., X_{n-1}),$$

then the source is said to be of *order* n. It is easy to show that

$$H[\mathbf{X}] = \lim_{n \to \infty} \frac{H(X_0, ...X_n)}{n+1}.$$

In general the outputs of the $X_j, j = 0, 1, ..., n$ are *dependent*. That is, the output of the communication process at step n depends on previous steps. Such serial correlation, in fact, is the very structure which enables most of what is done in this paper.

Here, however, the processes are all assumed stationary in time, that is, the serial correlations do not change in time, and the system is *stationary*.

A very broad class of such self-correlated, stationary, information sources, the so-called *ergodic* sources for which the long-run relative frequency of a sequence converges stochastically to the probability assigned to it, have a particularly interesting property:

It is possible, in the limit of large n, to divide all sequences of outputs of an ergodic information source into two distinct sets, S_1 and S_2, having, respectively, very high and very low probabilities of occurrence, with the source uncertainty providing the splitting criterion. In particular the Shannon-McMillan Theorem states that, for a (long) sequence having n (serially correlated) elements, the number of 'meaningful' sequences, $N(n)$ – those belonging to set S_1 – will satisfy the relation

$$\frac{\log[N(n)]}{n} \approx H[\mathbf{X}].$$

(8.2)

More formally,

$$\lim_{n \to \infty} \frac{\log[N(n)]}{n} = H[\mathbf{X}]$$

$$= \lim_{n \to \infty} H(X_n | X_0, ..., X_{n-1})$$

$$= \lim_{n \to \infty} \frac{H(X_0, ..., X_n)}{n+1}.$$

(8.3)

Using the internal structures of the information source permits *limiting attention only to high probability 'meaningful' sequences of symbols.*

8.2 The Rate Distortion Theorem

The Shannon-McMillan Theorem is the 'zero error limit' of the Rate Distortion Theorem (Dembo and Zeitouni, 1998; Cover and Thomas, 1991), which can be expressed in terms of a splitting criterion that identifies high probability pairs of sequences. We follow closely the treatment of Cover and Thomas (1991).

The origin of the problem is the question of representing one information source by a simpler one in such a way that the least information is lost. For example we might have a continuous variate between 0 and 100, and wish to represent it in terms of a small set of integers in a way that minimizes the inevitable distortion that process creates. Typically, for example, an analog audio signal will be replaced by a 'digital' one. The problem is to do this in a way which least distorts the *reconstructed* audio waveform.

Suppose the original stationary, ergodic information source Y with output from a particular alphabet generates sequences of the form

$$y^n = y_1, ..., y_n.$$

These are 'digitized,' in some sense, producing a chain of 'digitized values'

$$b^n = b_1, ..., b_n,$$

where the b-alphabet is much more restricted than the y-alphabet.

b^n is, in turn, *deterministically retranslated* into a reproduction of the original signal y^n. That is, each b^m is mapped on to a unique n-length y-sequence in the alphabet of the information source Y:

$$b^m \to \hat{y}^n = \hat{y}_1, ..., \hat{y}_n.$$

Note, however, that many y^n sequences may be mapped onto the *same* retranslation sequence $\hat{y}^n$, so that information will, in general, be lost.

The central problem is to explicitly minimize that loss.

The retranslation process defines a new stationary, ergodic information source, $\hat{Y}$.

The next step is to define a *distortion measure*, $d(y, \hat{y})$, which compares the original to the retranslated path. For example the *Hamming distortion* is

$$d(y, \hat{y}) = 1, y \neq \hat{y}$$

$$d(y, \hat{y}) = 0, y = \hat{y}.$$

(8.4)

For continuous variates the *Squared error distortion* is

$$d(y, \hat{y}) = (y - \hat{y})^2.$$

(8.5)

Possibilities abound.

The distortion between paths y^n and $\hat{y}^n$ is defined as

$$d(y^n, \hat{y}^n) = \frac{1}{n} \sum_{j=1}^{n} d(y_j, \hat{y}_j).$$

(8.6)

Suppose that with each path y^n and b^n-path retranslation into the y-language and denoted y^n, there are associated individual, joint, and conditional probability distributions

$$p(y^n), p(\hat{y}^n), p(y^n | \hat{y}^n).$$

The *average distortion* is defined as

$$D = \sum_{y^n} p(y^n)d(y^n, \hat{y}^n).$$

(8.7)

It is possible, using the distributions given above, to define the information transmitted from the incoming Y to the outgoing $\hat{Y}$ process in the usual manner, using the Shannon source uncertainty of the strings:

$$I(Y, \hat{Y}) \equiv H(Y) - H(Y|\hat{Y}) = H(Y) + H(\hat{Y}) - H(Y, \hat{Y}).$$

If there is no uncertainty in Y given the retranslation $\hat{Y}$, then no information is lost.

In general, this will not be true.

The *information rate distortion function* $R(D)$ for a source Y with a distortion measure $d(y, \hat{y})$ is defined as

$$R(D) = \min_{p(y,\hat{y}); \sum_{(y,\hat{y})} p(y)p(y|\hat{y})d(y,\hat{y}) \leq D} I(Y, \hat{Y}).$$

(8.8)

The minimization is over all conditional distributions $p(y|\hat{y})$ for which the joint distribution $p(y, \hat{y}) = p(y)p(y|\hat{y})$ satisfies the average distortion constraint (i.e., average distortion $\leq D$).

The Rate Distortion Theorem states that $R(D)$ is the minimum necessary rate of information transmission (effectively, the channel capacity) so that the average distortion does not exceed the distortion D. Cover and Thomas (1991) or Dembo and Zeitouni (1998) provide details.

Pairs of sequences $(y^n, \hat{y}^n)$ can be defined as *distortion typical*; that is, for a given average distortion D, defined in terms of a particular measure, pairs of sequences can be divided into two sets, a high probability one containing a relatively small number of (matched) pairs with $d(y^n, \hat{y}^n) \leq D$, and a low probability one containing most pairs. As $n \to \infty$, the smaller set approaches unit probability, and, for those pairs,

$$p(y^n) \geq p(\hat{y}^n|y^n) \exp[-nI(Y,\hat{Y})].$$

(8.9)

Thus, roughly speaking, $I(Y,\hat{Y})$ embodies the splitting criterion between high and low probability pairs of paths.

For the theory of interacting information sources, then, $I(Y,\hat{Y})$ can play the role of H in the dynamic treatment that follows.

The rate distortion function of equation (8.8) can actually be calculated in many cases by using a Lagrange multiplier method – see Section 13.7 of Cover and Thomas (1991). For a simple Gaussian channel having noise variance σ^2 then

$$R(D) = 1/2 \log[\sigma^2/D], 0 \leq D \leq \sigma^2,$$

$$R(D) = 0, D > \sigma^2.$$

(8.10)

For this particular channel, zero distortion, no mutations at all, requires an infinite channel capacity, which, the homology between information source uncertainty and free energy density implies, requires infinite energy.

A second important observation is that *any* rate distortion function $R(D)$, following the arguments of Cover and Thomas, (1991, Lemma 13.4.1) is necessarily a *decreasing convex function* of D, that is, a reverse-J-shaped curve. This requirement, like the singularity of Gaussian-like channels at zero distortion, has profound consequences for replication dynamics.

8.3 Morse Theory

Morse theory examines relations between analytic behavior of a function – the location and character of its critical points – and the underlying topology of the manifold on which the function is defined. We are interested in a number of such functions, for example information source uncertainty on a parameter space and 'second order' iterations involving parameter manifolds determining critical behavior, for example sudden onset of a giant component in the mean number model, and universality class tuning in the mean field model. These can be reformulated from a Morse theory perspective. Here we follow closely the elegant treatments of Pettini (2007) and Kastner (2006).

The essential idea of Morse theory is to examine an n-dimensional manifold M as decomposed into level sets of some function $f : M \to \mathbf{R}$ where $\mathbf{R}$ is the set of real numbers. The a-level set of f is defined as

$$f^{-1}(a) = \{x \in M : f(x) = a\},$$

the set of all points in M with $f(x) = a$. If M is compact, then the whole manifold can be decomposed into such slices in a canonical fashion between two limits, defined by the minimum and maximum of f on M. Let the part of M below a be defined as

$$M_a = f^{-1}(-\infty, a] = \{x \in M : f(x) \leq a\}.$$

These sets describe the whole manifold as a varies between the minimum and maximum of f.

Morse functions are defined as a particular set of smooth functions $f : M \to \mathbf{R}$ as follows. Suppose a function f has a critical point x_c, so that the derivative $df(x_c) = 0$, with critical value $f(x_c)$. Then f is a Morse function if its critical points are nondegenerate in the sense that the Hessian matrix of second derivatives at x_c, whose elements, in terms of local coordinates are

$$H_{i,j} = \partial^2 f / \partial x^i \partial x^j,$$

has rank n, which means that it has only nonzero eigenvalues, so that there are no lines or surfaces of critical points and, ultimately, critical points are isolated.

The index of the critical point is the number of negative eigenvalues of H at x_c.

A level set $f^{-1}(a)$ of f is called a critical level if a is a critical value of f, that is, if there is at least one critical point $x_c \in f^{-1}(a)$.

Again following Pettini (2007), the essential results of Morse theory are:

[1] If an interval $[a, b]$ contains no critical values of f, then the topology of $f^{-1}[a, v]$ does not change for any $v \in (a, b]$. Importantly, the result is valid even if f is not a Morse function, but only a smooth function.

[2] If the interval $[a, b]$ contains critical values, the topology of $f^{-1}[a, v]$ changes in a manner determined by the properties of the matrix H at the critical points.

[3] If $f : M \to \mathbf{R}$ is a Morse function, the set of all the critical points of f is a discrete subset of M, i.e. critical points are isolated. This is Sard's Theorem.

[4] If $f : M \to \mathbf{R}$ is a Morse function, with M compact, then on a finite interval $[a, b] \subset \mathbf{R}$, there is only a finite number of critical points p of f such that $f(p) \in [a, b]$. The set of critical values of f is a discrete set of $\mathbf{R}$.

[5] For any differentiable manifold M, the set of Morse functions on M is an open dense set in the set of real functions of M of differentiability class r for $0 \le r \le \infty$.

[6] Some topological invariants of M, that is, quantities that are the same for all the manifolds that have the same topology as M, can be estimated and sometimes computed exactly once all the critical points of f are known: Let the Morse numbers $\mu_i (i = 1, ..., m)$ of a function f on M be the number of critical points of f of index i, (the number of negative eigenvalues of H). The Euler characteristic of the complicated manifold M can be expressed as the alternating sum of the Morse numbers of any Morse function on M,

$$\chi = \sum_{i=0}^{m} (-1)^i \mu_i.$$

The Euler characteristic reduces, in the case of a simple polyhedron, to

$$\chi = V - E + F$$

where V, E, and F are the numbers of vertices, edges, and faces in the polyhedron.

[7] Another important theorem states that, if the interval $[a, b]$ contains a critical value of f with a single critical point x_c, then the topology of the set M_b defined above differs from that of M_a in a way which is determined by the index, i, of the critical point. Then M_b is homeomorphic to the manifold obtained from attaching to M_a an i-handle, i.e. the direct product of an i-disk and an $(m - i)$-disk.

Again, Pettini (2007) contains both mathematical details and further references. See, for example, Matusmoto (2002) or the classic by Milnor (1963).

8.4 Geodesic flows

Explicit parametization of $\mathcal{M}$ of Section 2.8 introduces standard – and quite considerable – notational complications (e.g. Burago et al., 2001; Auslander, 1967). Letting the parameters be a vector $\mathbf{K}$ having components $K_j, j = 1..m$, we can write $\mathcal{M}$ in terms of a 'metric tensor' $g_{i,j}(\mathbf{K})$ as

$$\mathcal{M}(A_0, A) = \int_A^{\hat{A}} [\sum_{i,j}^m g_{i,j}(\mathbf{K}) \frac{dK_i}{dt} \frac{dK_j}{dt}]^{1/2} dt$$

(8.11)

where the integral is taken over some parametized curve from the reference
state A_0 to some other state A. Then equation (2.12) becomes

$$[\sum_{i,j} g_{i,j}(\mathbf{K}) \frac{dK_i}{dt} \frac{dK_j}{dt}]^{1/2} = L\frac{dS}{d\mathcal{M}}.$$

(8.12)

This states that the 'velocity' $d\mathbf{K}/dt$ has a magnitude determined by the
local gradient in S at A_0, since the summation term on the left is the square
root of an inner product of a vector with itself.

The first condition of equation (2.13), i.e. setting $dS/d\mathcal{M}|_{A_0} = 0$, gives

$$\sum_{i,j} g_{i,j} \frac{dK_i}{dt} \frac{dK_j}{dt} = 0.$$

(8.13)

Thus the initial velocity is again zero, in the coordinates $\mathbf{K}$.

To go much beyond this obvious tautology we must, ultimately, generate
a parametized version of equation (8.11) and its dynamics, but expressing the
metric tensor, and hence redefining the geometry, in terms of derivatives of S
by the K_i. The result requires some development.

Write now, for parameters K_i the Onsager relation

$$dK_i/dt = L\partial S/\partial K_i,$$

(8.14)

where the K_i have been appropriately scaled.

Again place the system in a reference configuration A_0, having a vector of
parameters $\mathbf{K}_0$, so that

$$dK/dt|_{K_0} = L\nabla S|_{K_0} \equiv 0.$$

(8.15)

Deviations from this state, $\delta K \equiv K - K_0$, to first order, obey the relation

$$d\delta K_i/dt \approx L\sum_{j=1}^{m}(\frac{\partial^2 S}{\partial K_i \partial K_j}|_{K_0})\delta K_j.$$

(8.16)

In matrix form, writing $U_{i,j} = U_{j,i}$ for the partials in S, this becomes

$$d\delta\mathbf{K}/dt = LU\delta\mathbf{K}.$$

(8.17)

Assume appropriate regularity conditions on S and $\mathbf{U}$, and expand the deviations vector $\delta\mathbf{K}$ in terms of the m eigenvectors $\mathbf{e}_i$ of the symmetric matrix $\mathbf{U}$, having $\mathbf{U}\mathbf{e}_i = \lambda_i\mathbf{e}_i$, so that $\delta\mathbf{K} = \sum_{i=1}^{m}\delta a_i\mathbf{e}_i$.
Equation (8.17) then has the solution

$$\delta\mathbf{K}(t) = \sum_{i=1}^{m}\delta a_i\exp(L\lambda_i t)\mathbf{e}_i.$$

(8.18)

If all $\lambda_i \leq 0$, then the system is bounded quasistable, and a physiological forcing mechanism will be required to change status.
Next let $d\delta K_i/dt \equiv \delta V_i$. In first order the magnitude of the vector $\delta\mathbf{V}$ is

$$|\delta \mathbf{V}|^2 = \frac{L^2}{2} \sum_{i,j} [\sum_k U_{i,k} U_{k,j}] \delta K_i \delta K_j$$

(8.19)

Redefining

$$g_{i,j} \equiv \frac{L^2}{2} \sum_k U_{i,k} U_{k,j}$$

(8.20)

gives, after some notational shift, the symmetric Riemannian metric

$$dV^2 = \sum_{i,j} g_{i,j}(K_i, K_j) dK_i dK_j,$$

(8.21)

so that the metric, and hence the geometry, is now defined in terms of derivatives of S by the K_j.

The 'distance' between points a and b along some dynamic path in this geometry is, again,

$$s(A, B) = \int_A^B [\sum_{i,j} g_{i,j} \frac{dK_i}{dt} \frac{dK_j}{dt}]^{1/2} dt.$$

(8.22)

Application of the calculus of variations to minimize this expression produces a geodesic equation for the slowest dynamical path, and hence the most physiologically stable configuration. This has the traditional component-by-component form

$$d^2 K_i/dt^2 + \sum_{j,m} \Gamma^i_{j,m} \frac{dK_j}{dt} \frac{dK_m}{dt} = 0,$$

(8.23)

where the $\Gamma^i_{j,k}$ are the famous Christoffel symbols involving sums and products of $g_{i,j}$ and $\partial g_{i,j}/\partial K_m$ (e.g. Auslander, 1967; Burago et al., 2001; Wald, 1984, etc.).

The analog to equation (8.12) in this new geometry, defining a quasi-stable state, is that there exists a positive number $\mathcal{K} \ll R$, where R is the maximal possible number characteristic of the entire system, such that, for all geodesics $\mathbf{K}(t)$ which solve equation (8.23),

$$|\mathbf{K}(t)| \le \mathcal{K}.$$

(8.24)

at all times t.

Under such circumstances geodesics sufficiently near the reference state A_0 are all bound, and external physiological forcing must be imposed to cause a transition to a different condition. This result is analogous to the 'Black Hole' solution in General Relativity: recall that, within a critical radius near a sufficiently massive point source – the 'event horizon' – all geodesics, representing possible paths of light, are gravitationally bound without, however, the possible grace of some *deus ex machina*.

Note that the repulsive version of equation (8.24) might well be characterized as an unattainable White Hole.

Extending these considerations to the stochastic differential equation formulation of equation 3.15 leads quickly and decidedly into realms of stochastic differential geometry much like those of Emery (1989).

9

Mathematical Appendix II

9.1 Martingales

Suppose we have entered one of the great gambling casinos of the world, host to an almost infinite variety of games of chance: card games ranging from baccarat and blackjack to keno and poker, roulette wheels, one armed-bandits, dice games, and so on. Each game has different rules of play, even if, as for card games, the instruments of play are all the same. Complicated outcomes for those instruments produce equally complex patterns of loss or gain for the player.

Suppose a player begins with an initial fortune of some given amount, and bets $n = 1, 2, \ldots$ times according to a stochastic process in which a stochastic variable $\mathbf{X}_n$, which represents the size of the player's fortune at play n, takes values $\mathbf{X}_n = x_{n,i}$ with probabilities $P_{n,i}$ such that

$$\sum_i P_{n,i} = 1,$$

where i represents a particular outcome at step n.

Assume for all n there exists a value $0 < C < \infty$ such that the expectation of $\mathbf{X}_n$,

$$E(\mathbf{X}_n) \equiv \sum_i x_{n,i} P_{n,i} < C$$

(9.1)

for all n. That is, no infinite or endlessly increasing fortunes are permitted.

We note that the state $\mathbf{X}_n = 0$, having probability P_n^0, i.e. the loss of all a player's funds, terminates the game.

We suppose it possible to define conditional probabilities at step $n + 1$ which depend on the way in which the value of $\mathbf{X}_n$ was reached, so that we can define the conditional expectation of $\mathbf{X}_{n+1}$:

$$E(\mathbf{X}_{n+1}|\mathbf{X}_1, \mathbf{X}_2, ...\mathbf{X}_n) \equiv E(\mathbf{X}_{n+1}|n)$$

The 'sample space' for the probabilities defining this conditional expectation is the set of different possible sequences of the $x_{m,i} > 0$:

$$x_{1,i}, x_{2,j}, x_{3,k}...x_{n,q}$$

We call the sequence of stochastic variables $\mathbf{X}_n$ defining the game a *Submartingale* if, at each step n,

$$E(\mathbf{X}_{n+1}|n) \geq \mathbf{X}_n,$$

a *Martingale* if

$$E(\mathbf{X}_{n+1}|n) = \mathbf{X}_n$$

and a *Supermartingale* if

$$E(\mathbf{X}_{n+1}|n) \leq \mathbf{X}_n.$$

$\mathbf{X}_n$ is, remember, the player's fortune at step n.

Clearly a submartingale is favorable to the player, a martingale is an absolutely fair game, and a supermartingale is favorable to the house.

Regardless of the complexity of the game, the details of the playing instruments, the ways of determining gains or loss or their amounts, or any other structural factors of the underlying stochastic process, the essential content of the Martingale Limit Theorem is that in all three cases the sequence of stochastic variables $\mathbf{X}_n$ converges in probability 'almost everywhere' to a well-defined stochastic variable $\mathbf{X}$ as $n \rightarrow \infty$. That is, for each kind of martingale, no matter the actual sequence of winnings $x_{1,i}, x_{2,j}, ...x_{n,k}, x_{n+1,m}, ...$, you get to the same limiting stochastic variable $\mathbf{X}$. Sequences for which this does not happen have zero probability.

A simple proof of this result (Petersen, 1995) runs to several pages of dense mathematics using modern theories of abstract integration on sets. Indeed, all the asymptotic theorems we have cited require more or less arduous application of measure theory and Lebesgue integration, topics which are themselves relatively straightforward, elegant and worth study (Rudin, 1976, Royden, 1968). Proofs using more elementary approaches (Karlin and Taylor, 1975, Ch. 6) run to full chapters.

9.2 Nested Martingales

We are interested in a compound stochastic process in which the 'winnings' at the 'smaller' scale, played by one set of rules, contribute, in some sense, to a quite different game having completely different rules on a 'larger' scale. These games are bounded by the condition $E(\mathbf{X}_n) < C$, for some finite positive C.

The essential point is that a proportion of the winnings from the smaller game are duplicated by a 'benefactor' and directly raise the magnitude of the player's fortune for the larger, embedding game.

If the inner game is characterized at step n by the random variable $\mathbf{Y}_n$, then the 'real' winnings at step $n+1$ for the embedding game, associated with the random variable $\mathbf{X}_{n+1}$, become, for some function f_n, which may involve additional stochastic variables,

$$\mathbf{X}_{n+1} = f_n(\mathbf{X}_n, \mathbf{Y}_n, \mathbf{Y}_{n+1}).$$

(9.2)

A slightly different approach would involve conditional expectations in the convolution of scales:

$$E(\mathbf{X}_{n+1}|n) = F_n(\mathbf{X}_n, \mathbf{Y}_n, E(\mathbf{Y}_{n+1}|n))$$

(9.3)

for some function F_n.

Traditionally the simplest version of this extension assumes that the compound game is, in some sense, a subset of the original:

$$\mathbf{X}_{n+1} = \mathbf{X}_n + \mathbf{A}_n(\mathbf{Y}_{n+1} - \mathbf{Y}_n).$$

(9.4)

We assume the filter $\mathbf{A}_n \geq 0$ is a non-negative stochastic variable, which can indeed take the value 0. This may, for example, be greater than zero only one time in ten or a hundred, on average. Taking the conditional expectation gives

$$E(\mathbf{X}_{n+1}|n) = \mathbf{X}_n + \mathbf{A}_n(E(\mathbf{Y}_{n+1}|n) - \mathbf{Y}_n)$$

(9.5)

where we recognize the conditional expectation of any variate $\mathbf{Z}_n$ at step n is just its value.

Since $\mathbf{A}_n \geq 0$, *the game described by the attenuated sequence* $\mathbf{X}_n$ *has the same martingale classification as does the nested central city game described by* $\mathbf{Y}_n$.

9.3 The Martingale Transform

The X-processes in equation (9.4) is the *Martingale transform* of $\mathbf{Y}_n$ (Taylor, 1996, p.232; Billingsley, 1968, p. 412), and the result is classic, representing the *impossibility of a successful betting system*.

Note that the basic Martingale transform can be rewritten as

$$\frac{\mathbf{X}_{n+1} - \mathbf{X}_n}{\mathbf{Y}_{n+1} - \mathbf{Y}_n} \equiv \frac{\Delta\mathbf{X}_n}{\Delta\mathbf{Y}_n} = \mathbf{A}_n,$$

or

$$\Delta\mathbf{X}_n = \mathbf{A}_n\Delta\mathbf{Y}_n.$$

(9.6)

Induction gives

$$\mathbf{X}_{n+1} = \mathbf{X}_0 + \sum_{j=1}^{n} \mathbf{A}_j\Delta\mathbf{Y}_j.$$

(9.7)

This notation is suggestive: in fact the Martingale transform is the discrete analog of Ito's stochastic integral relative to a sequence of stopping times, (Taylor, 1996, p. 232; Protter, 1990, p. 44; Ikeda and Watanabe, 1989, p. 48). In the stochastic integral context the Y-process is called the 'integrator' and the A-process the 'integrand.' Further development leads toward generalizations of Brownian motion, the Poisson process, and so on (Meyer, 1989; Protter, 1990).

The basic picture is of the transmission of a signal, $\mathbf{Y}_n$, in the presence of noise, $\mathbf{A}_n$.

9.4 Stochastic Differential Equations

A more realistic extension of the elementary denumerable Martingale transform for our purposes is

$$\mathbf{X}_{n+1} = \mathbf{X}_n + (\mathbf{B}_{n+1} - \mathbf{B}_n)\mathbf{X}_n + \mathbf{A}_n(\mathbf{Y}_{n+1} - \mathbf{Y}_n),$$

(9.8)

where $\mathbf{B}_n$ is another stochastic variable.

Using the more suggestive notation of equations (9.6) and (9.7) this becomes the fundamental stochastic differential equation

$$\Delta\mathbf{X}_n = \mathbf{X}_n\Delta\mathbf{B}_n + \mathbf{A}_n\Delta\mathbf{Y}_n.$$

(9.9)

Taking conditional expectations gives

$$E(\mathbf{X}_{n+1}|n) - \mathbf{X}_n =$$

$$\mathbf{X}_n(E(\mathbf{B}_{n+1}|n) - \mathbf{B}_n) + \mathbf{A}_n(E(\mathbf{Y}_{n+1}|n) - \mathbf{Y}_n).$$

(9.10)

If $\mathbf{X}_n, \mathbf{A}_n \geq 0$, the martingale classification of $\mathbf{X}$ depends on those of $\mathbf{B}$ and $\mathbf{Y}$.

Extending the argument to a hierarchically-linked network is straightforward, leading to the Ito stochastic integral

$$\mathbf{X}_{n+1} \approx \mathbf{X}_0 + \sum_{k=1}^{n} \mathbf{A}_k \Delta \mathbf{Y}_k.$$

(9.11)

The complete hierarchical system, then undergoes an iterative Z-process defined by the integrator $\mathbf{X}_j$:

$$\mathbf{Z}_{m+1} \approx \mathbf{Z}_0 + \sum_{j=1}^{m} \mathbf{C}_j \Delta \mathbf{X}_j.$$

(9.12)

Extension of this development to intermediate times is complicated and involves taking the continuous limit of the Riemann-type sums of equations (9.7), (9.11) and (9.12). This produces the stochastic differential equation

$$d\mathbf{X}_t = \mathbf{X}_t d\mathbf{B}_t + \mathbf{A}_t d\mathbf{Y}_t$$

(9.13)

whose solution depends critically on the behavior of the second-order step-by-step 'quadratic variation,' a variance-like limit of the stochastic processes.

Letting $\mathbf{U}_n, \mathbf{V}_n$ be two arbitrary processes with $\mathbf{U}_0 = \mathbf{V}_0 = 0$, their quadratic variation is

$$[\mathbf{U}_n, \mathbf{V}_n] \equiv \sum_{j=1}^{n-1} (\mathbf{U}_{j+1} - \mathbf{U}_j)(\mathbf{V}_{j+1} - \mathbf{V}_j).$$

(9.14)

Taking the 'infinitesimal limit' of continuous time, a term-by-term expansion of this sum can be shown to give (e.g. Meyer, 1989; Protter, 1990)

$$[\mathbf{U}_t, \mathbf{V}_t] = \mathbf{U}_t \mathbf{V}_t - \int_0^t \mathbf{U}_s d\mathbf{V}_s - \int_0^t \mathbf{V}_r d\mathbf{U}_r.$$

(9.15)

To put this in some perspective, classical Brownian motion has the 'structure equation'

$$[\mathbf{X}_t, \mathbf{X}_t] = t.$$

That is, for Brownian motion the jump-by-jump quadratic variation increases linearly with time. While much of the contemporary theory of financial markets is based on Brownian analogs, real processes are likely to be more complex, subject to sudden, massive, discontinuous 'phase changes' which cannot be simply characterized as diffusional.

The solution of equation (9.13) is a classic result in the theory of stochastic differential equations (Protter, 1990). We assume for simplicity no discontinuous jumps, and first study the 'exponential' equation

$$d\mathbf{X}_t = \mathbf{X}_t d\mathbf{B}_t$$

or equivalently

$$\mathbf{X}_t = \mathbf{X}_0 + \int_0^t \mathbf{X}_s d\mathbf{B}_s$$

(9.16)

Following Protter (1990, p. 78) this has the solution

$$\mathbf{X}_t = \epsilon(\mathbf{B})_t = \mathbf{X}_0 \exp(\mathbf{B}_t - 1/2[\mathbf{B}_t, \mathbf{B}_t]).$$

(9.17)

Next we define

$$\mathbf{H}_t \equiv \int_0^t \mathbf{A}_s d\mathbf{Y}_s.$$

(9.18)

Equation (9.13) can be restated as

$$\mathbf{X}_t = \mathbf{H}_t + \mathbf{X}_0 + \int_0^t \mathbf{X}_s d\mathbf{B}_s.$$

(9.19)

For the continuous case, this has the formal solution (Protter, 1990, p.266)

$$\epsilon_{\mathbf{H}}(\mathbf{B})_t =$$

$$\epsilon(\mathbf{B})_t[\mathbf{H}_0 + \int_0^t 1/\epsilon(\mathbf{B})_s d(\mathbf{H}_s - [\mathbf{H}, \mathbf{B}]_s)],$$

with

$$1/\epsilon(\mathbf{B}) = \epsilon(-\mathbf{B} + [\mathbf{B}, \mathbf{B}]).$$

(9.20)

The structure equations defining $[\mathbf{B}, \mathbf{B}]$ and $[\mathbf{H}, \mathbf{B}]$ are critical in determining transient behavior, but not likely to have simple Brownian form.

10
References

Abler R., J. Adams and P. Gould, 1971, *Spatial Organization: The Geographer's View of the World*, Prentice-Hall, Englewood, NJ.

Acevedo-Garcia, D., 2000, Residential segregation and the epidemiology of infectious diseases, *Social Science and Medicine*, 51:1143-1161.

Adami C., and N. Cerf, 2000, Physical complexity of symbolic sequences, *Physica D*, 137:62-69.

Adami C., C. Ofria, and T. Collier, 2000, Evolution of biological complexity, *Proceedings of the National Academy of Sciences*, 97:4463-4468.

Agrawal, A., Q. Eastman, and D. Schatz, 1998, Implications of transposition mediated by V(D)J-recombination proteins RAG1 and RAG2 for origins of antigen-specific immunity, *Nature*, 394:744-751.

Aharony A., and D. Stauffer, 1984, Possible breakdown of the Alexander Orbach rule at low dimensionalities, *Physics Review Letters*, 52:2368-2373.

Aiello W., F. Chung, and L. Lu, 2000, A random graph model for massive graphs, in *Proceedings of the 32nd Annual ACM Symposium on the Theory of Computing*.

Albert R., and A. Barabasi, 2002, Statistical mechanics of complex networks, *Reviews of Modern Physics*, 74:47-97.

Alred D., 1998, Antigenic variation in Babesia Bovis: how similar is it to that of Plasmodium Falciparum? *Annals of Tropical Medicine and Parasitology*, 92:461-472.

Ancel L., 1999, A quantitative model of the Simpson-Baldwin effect, *Journal of Theoretical Biology*, 196:197-209.

Asanovic K, R. Bokik, B. Catanzaro, J. Gebis, P. Husbands, K. Keutzer, D. Patterson, W. Plishker, J. Shalf, S. Williams, and K. Yellick, 2006, The landscape of parallel computing research: a view from Berkeley, http://www.eecs.berkeley.edu/Pubs/TechRpts/2006/EECS-2006-183.pdf.

Ash, 1990, *Information Theory*, Dover Publications, New York.

Atlan H. and I. Cohen, 1998, Immune information, self-organization and meaning, *International Immunology*, 10:711-717.

Auslander L., 1967, *Differential Geometry*, Harper and Row, New York.

Avital E., and E. Jablonka, 2000, *Animal Traditions: Behavioral Inheritance in Evolution*, Cambridge University Press, New York.

Baars B., and S. Franklin, 2003, How conscious experience and working memory interact, *Trends in Cognitive Science*, doi:10.1016/S1364-6613(03)00056-1.

Baars B., 1988, *A Cognitive Theory of Consciousness*, Cambridge University Press, New York.

Baars, B., 2005, Global workspace theory of consciousness: toward a cognitive neuroscience of human experience, *Progress in Brain Research*, 150:45-53.

Bailey, N., 1975, *The Mathematical Theory of Infectious Diseases and its Applications*, Hafner, New York.

Baltimore, D., 2008, February, 2008, speech given at the annual meeting of the American Association for the Advancement of Science.

Bak A., R. Brown, G. Minian and T. Porter, 2006, Global actions, groupoid atlases and related topics, *Journal of Homotopy and Related Structures*, 1:1-54. Available from ArXiv depository.

Baker M., and J. Stock, 2007, Signal transduction: networks and integrated circuits in bacterial cognition, *Current Biology*, 17(4):R1021-4.

Barkow J., L. Cosmides and J. Tooby, eds., 1992, *The Adapted Mind: Biological Approaches to Mind and Culture*, University of Toronto Press.

Beck C. and F. Schlogl, 1995, *Thermodynamics of Chaotic Systems*, Cambridge University Press.

Bennett M., and P. Hacker, 2003 *Philosophical Foundations of Neuroscience*, Blackwell Publishing, London.

Bezemer D., S. Jurriaans, M. Prins, L. van der Hoek, J. Prins, F. de Wolr, B. Berkhout, R. Countinho, and N. Back, 2004, Declining trend in transmission of drug-resistant HIV-1 in Amsterdam, *AIDS*, 18:1571-1577.

Bignami G., 1982, Disease models and reductionist thinking in the biomedical sciences. In: S. Rose (ed.), *Against Biological Determinism*, pp 94-110, Allison and Busby, London.

Billingsley P, 1968, *Convergence of Probability Measures*, John Wiley and Sons, New York.

Binney J., N. Dowrick, A. Fisher, and M. Newman, 1986, *The Theory of Critical Phenomena*, Clarendon Press, Oxford, UK.

Bonner J., 1980, *The evolution of culture in animals*, Princeton University Press, Princeton, NJ.

Boyd, W., and M. Watts, 1997, Agro-industrial just in time: the chicken industry and postwar American capitalism. In D. Goodman, M. J. Watts (eds) *Globalising Food: Agrarian Questions and Global Restructuring*, Rutledge, London.

Brown R., 1987, From groups to groupoids: a brief survey, *Bulletin of the London Mathematical Society*, 19:113-134.

Brown S., and C. Getz, 2008, Towards domestic fair trade? Farm labor, food localsim, and the 'family scale' farm, *Geographic Journal*, 73:11-22.

Breitung, W., 2002, Transformation of a boundary regime: the Hong Kong and Mainland China case, *Environment and Planning A*, 34:1749-1762.

Burago D., Y. Burago, and S. Ivanov, 2001, *A Course in Metric Geometry*, American Mathematical Society, Providence, RI.

Burch, D., 2005, Production, consumption and trade in poultry: Corporate linkages and North-South supply chains. In N. Fold, W. Pritchard (eds.) *Cross-continental Food Chains*, Routledge, London.

Buxton Bridges, C., J. Katz, W. Seto, P. Chan, D. Tsang, et al., 2000, Risk of Influenza A (H5N1) Infection among Health Care Workers Exposed to Patients with Influenza A (H5N1), Hong Kong, *Journal of Infectious Disease*, 181:344-348.

Byrk A., and S. Raudenbusch, 2001, *Hierarchical Linear Models: Applications and Data Analysis Methods*, Sage Publications, New York.

Campos, P., and J. Fontanari, 1999, Finite-size scaling of the error threshold transition in finite populations, *Journal of Physics A*, 32:L1-L7.

Cannas Da Silva, A., and A. Weinstein, 1999, *Geometric Models for Noncommutative Algebras*, American Mathematical Society, New York.

Carter, C., and X. Li, 1999, Economic reform and the changing pattern of China's agricultural trade. Presented at the International Agricultural Trade Research Consortium, San Francisco, June 25-26, 1999.

Chiasson M., L. Berenson, W. Li, S. Schwartz, T. Singh, S. Forlenza, 1999, Declining HIV/AIDS mortality in New York City, *Journal of Acquired Immune Deficiency Syndrome*, 21:59-64.

Champagnat N., R. Ferriere, and S. Meleard, 2006, Univying evolutionary dynamics: From individual stochastic processes to macroscopic models, *Theoretical Population Biology*, 69:297-321.

Chang, W., 1969, National influenza experience in Hong Kong. *Bulletin of the World Health Organization*, 41:349-351.

Cheung, C., D. Vijaykrishna, G. Smith, X. Fan, J. Zhang et al., 2007, Establishment of influenza A virus (H6N1) in minor poultry in southern China, *Journal of Virology*, 81:10402-10412.

Clavel F., and A. Hance, 2004, HIV drug resistance, *New England Journal of Medicine*, 350:1023-1035.

Cliff, A., P. Haggett, J. Ord, 1986, *Spatial Aspects of Influenza Epidemics*, Pion, London.

Cohen I., 1992, The cognitive principle challenges clonal selection, *Immunology Today*, 13:441-444.

Cohen I., 2000, *Tending Adam's Garden: Evolving the Cognitive Immune Self*, Academic Press, New York.

Cohen I., 2006, Immune system computation and the immunological homunculus, in O. Nierstrasz et al. (eds), MoDELS 2006, *Lecture Notes in Computer Science*, 4199:499-512.

Cohen I., and D. Harel, 2007, Explaining a complex living system: dynamics, multi-scaling and emergence, *Journal of the Royal Society: Interface*, 4:175-182.

Combes C., 2000, Selective pressure in host-parasite systems [in French]. *J. Soc. Biol.*, 194:19-23.

Corless R., G. Gonnet, D. Hare, D. Jeffrey, and D. Knuth, 1996, On the Lambert W function, *Advances in Computational Mathematics*, 4:329-359.

Cosmides L. and J. Tooby, 1992, Cognitive adaptations for social exchange, in *The Adapted Mind: Evolutionary Psychology and the Generation of Culture*, Oxford University Press, New York.

Cover T., and J. Thomas, 1991, *Elements of Information Theory*, John Wiley Sons, New York.

Cristalli, A., and I. Capua, 2007, Practical problems in controlling H5N1 high pathogenicity avian influenza at village level in Vietnam and introduction of biosecurity measures, *Avian Disease*, 51(Suppl):461-462.

Crofton, H., 1971, Model of host-parasite relationships, *Parasitology*, 63:343-390.

Crosby A.W., 1986, *Ecological Imperialism: The Biological Expansion of Europe, 900-1900*, Cambridge University Press, Cambridge.

Davis, M., 2005 *The Monster at Our Door: The Global Threat of Avian Flu*, The New Press, New York.

Davis, M., 2006, *Planet of Slums*, Verso, London.

de Groot, J., M. Ruis, J. Scholten, J. Koolhaas and W. Boersma, 2001, Long-term effects of social stress on antiviral immunity in pigs, *Physiology and Behavior*, 73:145-158.

de Jong, M., C. Simmons, T. Thanh, V. Hien, G. Smith, et al., 2006, Fatal outcome of human influenza A (H5N1) is associated with high viral load and hypercytokinemia. *Nature Medicine*, 12:1203-1207.

Dembo A., and O. Zeitouni, 1998, *Large Deviations: Techniques and Applications*, Second edition, Springer, New York.

Dercole F., R. Ferriere, A. Gragnani and S. Rinaldi, 2006, Coevolution of slow-fast populations: evolutionary sliding, evolutionary pseudo-equilibria and complex Red Queen dynamics, *Proceedings of the Royal Society, B*, 273:983-990.

Diamond J., 1997, *Guns, Germs and Steel: the Fates of Human Societies*, W.W. Norton and Company, New York.

Diekmann U., and R. Law, 1996, The dynamical theory of coevolution: a derivation from stochastic ecological processes, *Journal of Mathematical Biology*, 34:579-612.

Dieckmann, U., J. Metz, M. Sabelis, and K. Sigmund, (eds.), 2002, *Adaptive Dynamics of Infectious Diseases: In Pursuit of Virulence Management*, Cambridge University Press, Cambridge, UK.

Dimitrov A., and J. Miller, 2001, Neural coding and decoding: communication channels and quantization, *Computation and Neural Systems*, 12:441-472.

DiNola J. and M. Neuberger, 2002, Altering the pathway of immunoglobin hypermutation by inhibiting uracil-DNA glycosylase, *Nature*, 419:43-48.

Dretske F., 1981, *Knowledge and the Flow of Information*, MIT Press, Cambridge, MA.

Dretske F., 1988, *Explaining Behavior*, MIT Press, Cambridge, MA.

Dretske F., 1992, What isn't wrong with folk psychology, *Metaphilosophy*, 29:1-13.

Dretske F., 1993, Mental events as structuring causes of behavior. In: (A. Mele and J. Heil, eds.), *Mental Causation*, pp. 121-136, Oxford University Press, Oxford.

Dretske F., 1994, The explanatory role of information, *Philosophical Transactions of the Royal Society A*, 349:59-70.

Duffy, G., O. Lyncha, and C. Cagneya, 2008, Tracking emerging zoonotic pathogens from farm to fork. Symposium on meat safety: from abattoir to consumer, *Meat Science*, 78:34-42.

Durham W., 1991, *Coevolution: Genes, Culture, and Human Diversity*, Stanford University Press, Palo Alto, CA.

Ebert, D., and J. Bull, 2008, The evolution and expression of virulence, in Stearns, S. and J. Koella (eds.), *Evolution in Health and Disease*, Second Edition, Oxford University Press, pp 153-167.

Eigen, M., 1971, Self-organization of matter and the evolution of biological macromolecules, *Naturwissenschaft*, 58:465-523.

Eigen, M., 1996, *Steps Towards Life: A Perspective on Evolution*, Oxford University Press, New York.

Eldredge N., and S. Gould, 1972, Punctuated equilibrium: an alternative to phyletic gradualism. In T. Schopf (ed.), *Models in Paleobiology*, 82-115, Freeman, Cooper and Co., San Francisco.

Eldredge N., 1985, *Time Frames: The Rethinking of Darwinian Evolution and the Theory of Punctuated Equilibria*, Simon and Schuster, New York.

Elenkov, I., and G. Chrousos, 1999, Stress, cytokine patterns and susceptibility to disease, *Bailliere's Clinical Endocrinology and Metabolism*, 13:583-595.

Emery M., 1989, *Stochastic Calculus in Manifolds*, Universitext series, Springer, New York.

English, T., 1996, Evaluation of evolutionary and genetic optimizers: no free lunch, int *Evolutionary Programming V: Proceedings of the Fifth Annual Conference on Evolutionary Programming*, L. Fogel, P. Angeline, and T. Back, eds., pp. 163-169. MIT Press, Cambridge, MA.

Erdos P., and A. Renyi, 1960, On the evolution of random graphs, reprinted in *The Art of Counting*, 1973, 574-618 and in *Selected Papers of Alfred Renyi*, 1976, 482-525.

Escoria M., L. Vazquez, S. Mendez, A. Rodriguez-Ropon, E. Lucio, and G. Nava, 2008, Avian influenza:genetic evolution under vaccination pressure, *Virology*, 5:15 doi:10.1186/1743-422X-5-15.

Ewald P., 2000, *Plague Time: How Stealth Infections Cause Cancers, Heart Disease, and Other Deadly Ailments*, The Free Press, New York.

Fan, C., 2001, Migration and labor-market returns in urban China: results from a recent survey in Guangzhou, *Environment and Planning A*, 33:479-508.

Fan, C., 2005, Interprovincial migration, population redistribution, and regional development in China: 1990 and 2000 census comparisons, *The Professional Geographer*, 57:295-311.

Fanon F., 1966, *The Wretched of the Earth*, Grove Press, New York.

Farmer, P., 2004, *Pathologies of Power: Health, Human Rights, and the New War on the Poor*, University of California Press, Berkeley.

Fasina, F., S. Bisschop, and R. Webster, 2007, Avian influenza H5N1 in Africa: an epidemiological twist, *Lancet Infectious Disease*, 7:696-697.

Fath B., H. Cabezas, and C. Pawlowski, 2003, Regime changes in ecological systems: an information theory approach, *Journal of Theoretical Biology*, 222:517-530.

Feller W., 1971, *An Introduction to Probability Theory and Its Applications*, John Wiley and Sons, New York.

Ferguson, N., 2007, Poverty, death, and a future influenza pandemic, *The Lancet*, 368:2187-2188.

Feynman R., 1996, *Feynman Lectures on Computation*, Addison-Wesley, Reading, MA.

Flegal K., M. Carroll, C. Ogden, and C. Johnson, 2002, Prevalence and trends in obesity among US adults, 1999-2000, *Journal of the American Medical Association*, 288:1723-1727.

Fleming, R., and C. Shoemaker, 1992, Evaluating models for spruce budworm-forest management: comparing output with regional field data, *Ecological Applications*, 2:460-477.

Fontanari J., M. Santos, and E. Szathmary, 2006, Coexistence and error propagation in pre-biotic vesicle models: a group selection method, *Journal of Theoretical Biology*, 239:247-256.

Food and Agriculture Organization of the United Nations, 2004, Questions and answers on avian influenza; briefing paper prepared by AI Task Force. Internal FAO document, January 30.
http://www.animal-health-online.de/drms/faoinfluenza.pdf.

Food and Agriculture Organization of the United Nations, 2003, *World Agriculture: Towards 2015/2030: An FAO Perspective*, Earthscan Publications, London.

Forlenza M., and A. Baum, 2000, Psychosocial influences on cancer progression: alternative cellular and molecular mechanisms, *Current Opinion in Psychiatry*, 13:639-645.

Franzosi R., and M. Pettini, 2004, Theorem on the origin of phase transitions, *Physical Review Letters*, 92:060601.

Fredlin M., and A. Wentzell, 1998, *Random Perturbations of Dynamical Systems*, Springer, New York.

Freeman H. and C. McCord, 1990, Excess mortality in Harlem, *New England Journal of Medicine*, 322:173-180.

French, H., 2006, Wealth grows, but health care withers in China, *New York Times*, January 14, 2006.

Garrett K, and C. Cox, 2008, Applied biodiversity science: managing emerging diseases in agriculture and linked natural systems using ecological principles. In Ostfield, R., F. Keesing, and V. Eviner (eds.), *Infectious Disease Ecology: Effects of Ecosystems on Disease and of Disease on Ecosystems*, Princeton University Press, Princeton, NJ.

Garrett, L., 2001, *Betrayal of Trust: The Collapse of Global Public Health*, Oxford University Press, Oxford, UK.

Gavrilets, S., 2003, Models of speciation: what have we learned in 40 years?, *Evolution*, 57:2197-2215.

Gearhart P., 2002, The roots of antibody diversity. *Nature*, 419:29-31.

Gibson, G., and I. Dworkin, 2004, Uncovering cryptic genetic variations, *Nature Reviews Genetics*, 5:681-691.

Gilbert, M., P. Chaitaweesub, T. Parakamawongsa, S. Premashthira, T. Tiensin, et al., 2006, Free-grazing ducks and highly pathogenic avian influenza, Thailand, *Emerging Infectious Diseases*, 12:227-234.

Gilbert M., X. Xiao, D. Pfeiffer, M. Epprecht, S. Boles, et al., 2008, Mapping H5N1 highly pathogenic avian influenza risk in Southeast Asia, *Proceedings of the National Academy of Sciences*, 105:4769-4774.

Gilbert M., X. Xiao, W. Wint, and J. Slingenbergh, 2007, Poultry production dynamics, bird migration cycles, and the emergence of highly pathogenic avian influenza in East and Southeast Asia, in: Sauerborn, R., L. Valrie, (eds.), *Global Environmental Change and Infectious Diseases: Impacts and Adaption Strategies*, Springer Verlag, Berlin.

Gilbert M., A. Rambaut, G. Wlasiuk, T. Spira, A. Pitchenik and M, Worobey, 2007, The emergence of HIV/AIDS in the Americas and beyond, *Proceedings of the National Academy of Sciences*, 104:18566-18570.

Gilbert, S., 2000, Diachronic biology meets evo-devo: C.H. Waddington's approach to evolutionary developmental biology, *American Zoologist*, 40:729-737.

Gilbert, S., 2001, Mechanisms for the environmental regulation of gene expression: ecological aspects of animal development, *Journal of Bioscience*, 30:65-74.

Gilbert, S., 2005, Ecological developmental biology: developmental biology meets the real world, *Developmental Biology*, 233:1-12.

Glaser, R., J. Kiecolt-Glaser, W. Malarkey and J. Sheridan, 1998, The influence of psychological stress on the immune system response to vaccines, *Annals of the New York Academy of Sciences*, 840:649-655.

Glaser, R., J. Sheridan, W. Malarkey, R. MacCallum and K. Kiecolt-Glaser, 2000, Chronic stress modulates the immune response to a pneumococcal pneumonia vaccine, *Psychosomatic Medicine*, 62:804-807.

Glazebrook J., and R. Wallace, 2007, Rate distortion manifolds as model spaces for cognitive information. Submitted.

Golubitsky M., and I. Stewart, 2006, Nonlinear dynamics and networks: the groupoid formalism, *Bulletin of the American Mathematical Society*, 43:305-364.

Gordon A., 2000, Cultural identity and illness: Fulani views, Culture, *Medicine and Psychiatry*, 24:297-330.

Gordon, S., I. Pandrea, R. Dunham, C. Apetvel, and G. Silvestri, 2005, The call of the wild: what can be learned from studies of SIV infection of natural hosts? Theoretical Biology and Biophysics Group, LANL.

Goubault, E., and M. Raussen, 2002, Dihomotopty as a tool in state space analysis, *Lecture Notes in Computer Science*, Vol. 2286, April, 2002, pp. 16-37, Springer, New York.

Goubault, E., 2003, Some geometric perspectives in concurrency theory, *Homology, Homotopy, and Applications*, 5:95-136.

Gould P. and G. Tornqvist, 1971, Information, innovation and acceptance, in T. Hagerstrand and A Kuklinski (eds.), *Information Systems for Regional Development*, Lund series in Geography, Ser. B, Human Geography, No. 37, 149-168.

Gould, P., 1993, *The Slow Plague*, Blackwell, Oxford.

Gould, P., 1999, *Becoming a Geographer*, Syracuse University Press.

Gould, S., 2002, *The Structure of Evolutionary Theory*, Harvard University Press, Cambridge, MA.

Gould, S., 2003, *The Hedgehog, the Fox, and the Magister's Pox*, Harmony Books, New York.

Gould, S., and N. Eldredge, 1977, Punctuated equilibrium: the tempo and mode of evolution reconsidered, *Paleobiology*, 3:115-151.

Grant R., F. Hecht, M. Warmerdam, et al., 2002, Time trends in primary HIV-1 drug resistance among recently infected persons, *Journal of the American Medical Association*, 288:181-188.

Granovetter M., 1973, The strength of weak ties, *American Journal of Sociology*, 78:1360-1380.

Greger, M., 2006, *Bird Flu: A Virus of our own Hatching*, Latern Books, New York.

Grimmett G., and A. Stacey, 1998, Critical probabilities for site and bond percolation models, *The Annals of Probability*, 4:1788-1812.

Grossman Z., 1989, The concept of idiotypic network: deficient or premature? In: H. Atlan and I.R. Cohen (eds.), *Theories of Immune Networks*, Springer Verlag, Berlin, p. 3852.

Grossman, Z., 2000, Round 3, *Seminars in Immunology*, 12:313-318.

Gryazeva, N., A. Shurlygina, L. Verbitskaya, E. Mel'nikova, N. Kudryavtseva and V. Trufakin, 2001, Changes in various measures of immune status in mice subject to chronic social conflict, *Neuroscience and Behavioral Physiology*, 31:75-81.

Gu C., J. Shen, W. Kwan-Yiu, and F. Zhen, 2001, Regional polarization under the socialist-market system since 1978: a case study of Guangdong province in South China, *Envionment and Planning A*, 33:97-119.

Guan Y., K, Shortridge, S. Krauss, and R. Webster, 1999, Molecular characterization of H9N2 influenza viruses: were they the donors of the 'internal' genes of H5N1 virusus in Hong Kong? *Proceedings of the National Academy of Sciences*, 96:9363-9367.

Guldin, G., 1993, Urbanizing the countryside: Guangzhou, Hong Kong and the Pearl River Delta, in Guldin, G., (ed.) *Urbanizing China*, Greenwood Press, Westport, CT., pp. 157-184.

Gunderson L., 2000, Ecological resilience – in theory and application, *Annual Reviews of Ecological Systematics*, 31:425-439.

Gunderson L., 2007, Personal communication.

Haley, G., C. Tan, and U. Haley, 1998, *The New Asian Emperors: The Chinese Overseas, their Stategies and Competitive Advantages*, Butterworth Heinemann, London.

Hammond, E., 2007, Flu virus sharing summit: wrap-up. 'Effect Measure' blog.

http://scienceblogs.com/effectmeasure/2007/11/flu-virus-sharing-summit-wrap-1.php.

Hammond, E., 2008, Material Transfer Agreement hypocrisy. 'Immunocompetent' blog., Posted August 11, 2008.

Hartl D., and A. Clark, 1997, *Principles of Population Genetics*, Sinaur Associates, Sunderland, MA.

Hart-Landsberg, M., and P. Burkett, 2005a, *China and Socialism: Market Reforms and Class Struggle*, Monthly Review Press, New York.

Hart-Landsberg, M., and P. Burkett, 2005b, China and socialism: engaging the issues, *Critical Asian Studies*, 37:597-628.

Harvey, D., 1982/2006, *The Limits to Capital*, Verso Press, New York.

Heartfield, J., 2005, China's comprador capitalism is coming home, *Review of Radical Political Economics*, 37:196-214.

Hertel, T., A. Nin-Pratt, A. Rae, and S. Ehui, 1999, Productivity growth and 'catching-up': implications for China's trade in livestock products. Paper for presentation at the International Agicultural Trade Research Consortium meeting on China's Agricultural Trade Policy, San Francisco, CA, June 25-26.

Hertel, T., K. Anderson, J. Francois, and W. Martin, 2000, Agriculture and non-agricultural liberalization in the Millennium Round. Policy Discussion Paper No. 0016, Centre for International Economic Studies, University of Adelaide.

https://www.gtap.agecon.purdue.edu/resources/download/689.pdf.

Hodgson G., 1993, *Economics and Evolution: Bringing Life Back Into Economics*, University of Michigan Press, Ann Arbor, MI.

Hoffmann, E., J. Stech, I. Leneva, S. Krauss, C. Scholtisek, et al., 2000, Characterization of the influenza A virus gene pool in avian species in Southern China: was H6N1 a derivative or a percursor of H5N1?, *Journal of Virology*, 74:6309-6315.

Holling C., 1973, 1973, Resilience and stability of ecological systems, *Annual Reviews of Ecological Systematics*, 4:1-23.

Holling C., 1992, Cross-scale morphology, geometry and dynamics of ecosystems, *Ecological Monographs*, 41:1-50.

Holmes, E., 2003, Error thresholds and the constraints to RNA virus evolution, *TRENDS in Microbiology*, 11:543-546.

Holmes, E., 2005, On being the right size, *Nature Genectics*, 37:923-924.

Holmes, E., 2006, The evolution of viral emergence, *Proceedings of the National Academy of Sicences*, 103:4803-4804.

Hughes, J.D., 2001, *An Environmental History of the World: Humankind's Changing Role in the Community of Life*, Routledge, London.

Ikeda N., and S. Watanabe, 1989, *Stochastic Differential Equations and Diffusion Processes*, 2nd Edition, North Holland Publishing Co., Amsterdam.

Jablonka, E., 2001, The systems of inheritance, in S. Oyama, P. Griffiths and R. Gray (eds.), *Cycles of Contingency: Developmental Systems and Evolution*, The MIT Press, Cambridge, MA.

Jeffries, R., R. Rockwell, and K. Abraham, 2004, The embarrassment of riches: agricultural food subsidies, high goose numbers, and loss of Artic wetlands-a continuing saga, *Environment Review*, 11:193-232.

Jimenez-Montano M., 1989, Formal languages and theoretical molecular biology, in *Theoretical Biology: Epigenetic an Evolutionary Order in Complex Systems*, B. Goodwin and P. Saunders (eds.), Edinburgh University Press.

Johnson, G., 1992, The political economy of Chinese urbanization: Guangdong and the Pearl River Delta region. In Guldin, G. (ed.), *Urbanizing China*, Greenwood Press, Westpor, CT.

Jones K., N. Patel, M. Levy, A. Storeygard, D. Balk, J. Gittleman, and S. Daszak, Global trends in emerging infection, *Nature*, 451:990-993.

Kandun, I., H. Wibisono, E. Sedyaningsih, Yusharmen, W. Hadisoedarsuno, et al., 2006, Three Indonesian clusters of H5N1 virus infection in 2005, *New England Journal of Medicine*, 355:2186-2194.

Kang-Chung, N., 1997, Chicken imports slashed by third, *South China Morning Post*, December, 15.

Karlin S., and H. Taylor, 1975, *A First Course in Stochastic Processes*, second edition, Academic Press, New York.

Kastner M., 2006, Phase transitions and configuration space topology, arXiv preprint cond-mat/0703401

Khinchin A., 1957, *Mathematical Foundations of Information Theory*, Dover Publications, New York.

Kiecolt-Glaser, K., R. Glaser, S. Gravenstein, W. Malarkey and J Sheridan, 1996, Chronic stress alters the immune response to influenza virus vaccine in older adults, *Proceedings of the National Academy of Sciences* 93:3043-3047.

Kilpatrick, A., A. Chmura, D. Gibbons, R. Fleischer, P. Marra, and P. Daszak, 2006, Predicting the global spread of H5N1 avian influenza, *Proceedings of the National Academy of Sciences*, 103:19368-19373.

Kim, J., J. Millen, A. Irwin, and J. Gershman, (eds.), 2000, *Dying for Growth: Global Inequality and the Health of the Poor*, Common Courage Press, Boston.

Kozma R., M. Puljic, P. Balister, B. Bollobas, and W. Freeman, 2004, Neuropercolation: a random cellular automata approach to spatio-temporal neurodynamics, *Lecture Notes in Computer Science*, 3305:435-443.

Kozma R., M. Puljic, P. Balister, and B. Bollobas, 2005, Phase transitions in the neuropercolation model of neural populations with mixed local and non-local interactions, *Biological Cybernetics*, 92:367-379.

Krebs, P., 2005, Models of cognition: neurological possibility does not indicate neurological plausibility, in Bara, B., L. Barsalou, and M. Bucciarelli (eds.), *Proceedings of CogSci 2005*, pp. 1184-1189, Stresa, Italy. Available at http://cogprints.org/4498/.

Laland K., F. Odling-Smee, and M. Feldman, 1999, Evolutionary consequences of niche construction and their implications for ecology, *Proceedings of the National Academy of Sciences*, 96:10242-10247.

Landau L., and E. Lifshitz, 2007, *Statistical Physics*, 3rd edition, London, Pergamon.

Lee J., 2000, *Introduction to Topological Manifolds*, Springer, New York.

Leff, B., N. Ramankutty, and J. Foley, 2004, Geographic distribution of major crops across the world, *Global Biogeochemical Cycles*, 18:GB1009, doi:1029/2003GB002108.

Lemly, A., R. Kingsford , and J. Thompson, 2000, Irrigated agriculture and wildlife conservation: conflict on a global scale, *Environmental Management*, 25:485-512.

Leuthausser, I., 1986, An exact correspondence between Eigen's evolution model and a two-dimensional Ising system, Journal of Chemical Physics, 84:1884-1885.

Levin S., 1989, Ecology in theory and application, in *Applied Mathematical Ecology*, S. Levin, T. Hallam, and L. Gross (eds.), Biomathematics Texts 18, Springer-Verlag, New York.

Levin, S., 1999, Towards a science of ecological management, *Ecology and Society*, 3(2):6.

Levins, R., 1993, The ecological transformation of Cuba, *Agriculture and Human Values*, 10:52-60.

Levins R., 1998, The internal and external in explanatory theories. *Science as Culture*, 7:557-582.

Levins, R., 2007, How Cuba is going ecological, in R. Lewontin and R. Levins, *Biology Under the Influence: Dialectical Essays on Ecology, Agriculture, and Health*, Monthly Review Press, New York.

Levins R., and R. Lewontin, 1985, *The Dialectical Biologist*, Harvard University Press, Cambridge MA.

Lewontin R., 1993, *Biology as Ideology: The Doctrine of DNA*, Harper Collins, New York.

Lewontin R., 2000, *The Triple Helix: gene, organism, and environment*, Harvard University Press.

Lewontin, R., 2007, The maturing of capitalist agriculture: farmer as proletarian. In Lewontin, R., and R. Levins, *Biology Under the Influence: Dialec-*

tical Essays on Ecology, Agriculture, and Health. Monthly Review Press, New York.

Ley, R., C. Lozupone, P. Tumbagh, et al., 2008, Evoalution of mammals and their gut flora, *Science*, 230:1647-1651.

Li, K., Y. Guan, J. Wang, G. Smith, K. Xu, et al., 2004, Genesis of a highly pathogenic and potentially pandemic H5N1 influenza virus in eastern Asia, *Nature*, 430: 209-213.

Li, M., 2008, An age of transition: the United States, China, Peak Oil, and the demise of neoliberalism, *Monthly Review*, 59:20-34.

Liao J., R. Biscolo, Y. Yang, L. My Tran, C. Sabatti, and V. Roychowdhury, 2003, Network component analysis: Reconstruction of regulatory signals in biological systems, *Proceedings of the National Academy of Sciences*, 100:15522-15527.

Lin, G., 1997, *Red Capitalism in South China: Growth and Development of the Pearl River Delta*, UBC Press, Vancouver.

Lin, G., 2000, State, capital, and space in China in an age of volatile globalization, *Environment and Planning A*, 32:455-471.

Lipsitch, M., and M. Nowak, 1995, The evolution of virulence in sexually transmitted HIV/AIDS, *Journal of Theoretical Biology*, 174:427-440.

Liu, J., K. Okazaki, W. Shi, Q. Wu, A. Mweene, and H. Kida, 2003, Phylogenetic analysis of neuraminidase gene of H9N2 influenza viruses prevalent in chickens in China during 1995-2002, *Virus Genes*, 27:197-202.

Liu, Y., and M. Ringner, 2007, Revealing signaling pathway deregulation by using gene expression signatures and regulatory motif analysis, *Genome Biology*, 8:R77.

Luchinsky D., 1997, On the nature of large fluctuations in equilibrium systems: observations of an optimal force, *Journal of Physics A*, 30:L577-L583.

Luczak T., 1990, *Random Structures and Algorithms*, 1:287.

Luo, X., Y. Ou, and X. Zhou, 2003, Livestock and poultry production in China. Presented at Bioproduction in East Asia: Technology Development Globalization Impact, a pre-conference forum in conjunction with the 2003 ASAE Annual International Meeting, 27 July 2003, (Las Vegas, Nevada, USA) ASAE Publication Number 03BEA-06, ed. Chi Thai.

Manning, L., and R. Baines, 2004, Globalization: a study of the populatry-meat supply chain, *British Food Journal*, 106:819-836.

Manning, L., R. Baines, and S. Chadd, 2007, Trends in global poultry meat supply chain, *British Food Journal*, 109:332-342

Marx, K., 1867/1990, *Capital: A Critique of Political Economy*, Volume 1, Penguin Group, London.

Massey D. and N. Denton, 1993, *American Apartheid: Segregation and the Making of the Underclass*, Harvard University Press, Cambridge, MA.

Matsumoto Y., 2002, *An Introduction to Morse Theory*, Translations of Mathematical Monographs, Vol. 208, American Mathematical Society.

Mayr E., 1996, The autonomy of biology: the position of biology among the sciences. *The Quarterly Review of Biology*, 71:97-106.

McCauley J, 1993, *Chaos, Dynamics and Fractals: An Algorithmic Approach to Deterministic Chaos*, Cambridge Nonlinear Science Series, Cambridge, UK.

McMichael, P., 2006, Feeding the world: agriculture, development, and ecology. In Panitch, L., and C. Leys (eds.), *Socvialist Register 2007: Coming to Terms with Nature*, Merlin Press, London.

Meyer P., 1989 A short presentation of stochastic calculus, appendix to *Stochastic Calculus on Manifolds*, M. Emery, Springer, New York.

Michel L., and J. Mozrymas, 1977, Application of Morse Theory to the symmetry breaking in the Landau theory of the second order phase transition, in *Group Theoretical Methods in Physics: Sixth International Colloquium*, eds. P. Kramer, A. Rieckers, Lecture Notes in Physics, Vol. 79, pp. 447-461, SPringer, New York.

Milnor J., 1963, *Morse Theory*, Annals of Mathematical Studies, Vol. 51, Princeton University Press.

Modiano D., V. Petrarca, B. Sirma, I. Nebie, D. Diallo, F. Esposito and M. Coluzzi, 1996, Different response to *Plasmodium falciparum* malaria in West African sympatric ethnic groups, *Proceedings of the National Academy of Sciences* 93:13206-13211.

Modiano D., G. Luoni, V. Petrarca, B. Sodiomon Sirima, M. De Luca, J. Simpore, M. Coluzzi, J. Bodmer and G. Modiano, 2001 HLA class I in three West African ethnic groups: genetic distances from sub-Saharan and Caucasoid populations, *Tissue Antigen* 57:128-137.

Modiano D., G. Luoni, B. Sirima, A. Lanfrancotti, V. Petrarca, F. Cruciani, J. Simpore, B. Ciminelli, E. Foglietta, P. Grisanti, I. Bianco, G. Modiano and M. Coluzzi, 2001, The lower susceptibility to Plasmodium falciparum malaria of Fulani of Burkina Faso (west Africa) is associated with low frequencies of classic malaria-resistance genes, *Transactions of the Royal Society of Tropical Hygiene and Medicine*, 95:149-152.

Modiano D., V. Petrarca, B. Sirima, I. Nebie, G. Luoni, F. Esposito and M. Coluzzi, 1998, Baseline immunity of the population and impact of insecticide-treated curtains on malaria infection, *American Journal of Tropical and Medical Hygiene*, 59:336-340.

Molle, F., 2007, Scales and power in river basin management: the Chao Phraya River in Thailand, *The Geographical Journal*, 173:358-373.

Molloy M., and B. Reed, 1995, A critical point for random graphs with a given degree sequence, *Random Structures and Algorithms*, 6:161-179.

Molloy M., and B. Reed, 1998, The size of the giant component of a random graph with a given degree sequence, *Combinatorics, Probability, and Computing*, 7:295-305.

Mukhtar, M., S. Rasool, D. Song, C. Zhu, O. Hao, et al., 2007, Origins of highly pathogenic H5N1 avian influenza virus in China and genetic characterization of donor and recipient viruses, *Journal of Gneeral Virology*, 88:3094-3099.

Myers, K., S. Setterquist, A. Capuano, and G. Gray, 2007, Infection due to 3 avian influenza subtypes in United States veterinarians, *Clinical Infectious Disease*, 45:4-9.

Newman K., and E. Wyly, 2006, The right to stay put, revisited: gentrification and resistance to displacement in New York City, *Urban Studies*, 43:23-57.

Newman M., S. Strogatz, and D. Watts, 2001, Random graphs with arbitrary degree distributions and their applications, *Physical Review E*, 64:026118, 1-17.

Newman M., 2003, Properties of highly clustered networks, arXiv:cond-mat/0303183v1.

Nisbett R., K. Peng, C. Incheol and A. Norenzayan, 2001. Culture and systems of thought: holistic vs. analytic cognition. *Psychological Review*, 108:291-310.

Nowak M. and R. May, 2000. *Virus Dynamics: the Mathematical Foundations of Immunology and Virology*, Oxford University Press, New York.

Nunney L., 1999, Lineage selection and the evolution of multistage carcinogenesis, *Proceedings of the London Royal Society B*, 266:493-498.

NYCDOH, 2006, HIV epidemiology program, 2nd semiannual report, HIV Epidemiology Program, 346 Broadway, Rm. 706, CN44, New York, NY, 10013, http://www.nyc.gov/html/doh/html/dires/hivepi.shtml

Odling-Smee, F., K. Laland, and M. Feldman, 1996, Niche construction, *The American Naturalist*, 147:641-648.

Odling-Smee, F., K. Laland, and M. Feldman, 2003, *Niche Construction: The Neglected Process in Evolution*, Princeton University Press, Princeton, NJ.

Ofria C., C. Adami, and T. Collier, 2003, Selective pressures on genomes in molecular evolution, *Journal of Theoretical Biology*, 222:477-483.

Ogata T., Y. Yamazaki, N. Okabe, Y. Nakamura, M. Tashiro, et al., 2008, Human H5N2 avian influenza infection in Japan and the factors associated with high H5N2-neutralizing antibody titer, *Journal of Epidemiology*, Jul 7. [Epub ahead of print].

Onsager L., and S. Machlup, 1953, Fluctuations and irreversible processes, *Physical Review*, 91:1505-1512.

O'Nuallain S., 2006, Context in computational linguistics and gene expression, http://www-clsi.stanford.edu/events/Coglunch/nuallain-2006.

OECD, 1998, *Agricultural Policies in Non-Member Countries*, Centre for Cooperation with Economies in Transition, OECD, Paris.

Oyama S., P.E. Griffiths, and R.D. Gray (eds.), 2001, *Cycles of Contingency: Developmental Systems and Evolution*, The MIT Press, Cambridge, MA.

Perkins, F., 1997, Export performance and enterprise reform in China's coastal provinces, *Economic Development and Cultural Change*, 45:501-539.

Pettini M., 2007, *Geometry and Topology in Hamiltonian Dynamics and Statistical Mechanics*, Springer, New York.

Phongpaichit, P., and C. Baker, 1995, *Thailand, Economy and Politics*, Oxford University Press, Oxford, UK.

Pielou E.C., 1977, *Mathematical Ecology*, John Wiley and Sons, New York.

Petersen K., 1995, *Ergodic Theory*, Cambridge Studies in Advanced Mathematics 2, Cambridge University Press, Cambridge, UK.

Pettini, M., 2007, *Geometry and Topology in Hamiltonian Dynamics and Statistical Mechanics*, Springer, New York.

Podolsky S., and A. Tauber, 1998, *The Generation of Diversity: Clonal Selection Theory and the Rise of Molecular Biology*, Harvard University Press.

Pohl W., 1962, Differential geometry of higher order, *Topology* 1:169-211.

Poon, L., Y. Guan, J. Nicholls, K. Yuen, and J. Peiris, 2004, The aetiology, origins, and diagnosis of severe acute respiratory syndrome, *Lancet Infectious Disease*, 4:663-671.

Pratt V., 1991, Modeling concurrency with geometry, *Proceedings of the 18th ACM SIGPLAN-SIGACT Symposium on Principles of Programming Languages*, 311-322.

Priami C., 2007, Computational thinking in biology, *Transactions on Computational Systems Biology*, VIII, LNBI 4780, 63-76.

Protter P., 1990, *Stochastic Integration and Differential Equations: A New Approach*, Springer-Verlag, New York.

Puzelli S., L. Di Trani, C. Fabiani, L. Campitelli, M. De Marco, et al., 2005, Serological analysis of serum samples from humans exposed to avian H7 influenza viruses in Italy between 1999 and 2003, *Journal of Infectious Disease*, 192:1318-1322.

Pyle G., 1969, Diffusion of cholera in the United States, *Geographical Analysis*, 1:59-75.

Rambaut A., D. Posada, K. Crandall and E. Holmes, 2004, The causes and consequences of HIV evolution, *Nature Reviews/Genetics*, 5:52-61.

Richerson P. and R. Boyd, 1995, The evolution of human hypersociality, Paper for Rindberg Castle Symposium on Ideology, Warfare, and Indoctrination, (January, 1995), and HBES meeting, 1995.

Richerson P., and R. Boyd, 2004, *Not by Genes Alone: How Culture Transformed Human Evolution*, Chicago University Press.

Ricotta C., 2003, Additive partition of parametric information and its associated β-diversity measure, *Acta Biotheoretica*, 51:91-100.

Ridley M., 1996, *Evolution*, Second Edition, Blackwell Science, Oxford University Press.

Rohowsky-Kochan, C., J. Skurnick, D. Molinaro and D. Louria, 1998, HLA antigens associated with susceptibility/resistance to HIV-1 infection, *Human Immunology*, 59:802-815.

Rojdestvensky I. and M. Cottam, 2000, Mapping of statistical physics to information theory with applications to biological systems, *Journal of Theoretical Biology*, 202:43-54.

Royden H., 1968, *Real Analysis*, Macmillan, New York.

Rozelle, S., C. Pray, and J. Huang, 1999, Importing the means of production: foreign capital and technologies flows in China's agriculture. Paper presented at the 1999 IATRC Conference, San Francisco, CA, June 25-26. http://www.agecon.ucdavis.edu/people/faculty/facultydocs/Sumner /iatrc/rozelle.pdf

Rudin W., 1976,*Principles of Mathematical Analysis*, McGraw-Hill, New York.

Rweyemamu, M., R. Paskin, A. Benkirane, V. Martin, P. Roeder, et al., 2000, Emerging diseases of Africa and the Middle East, *Annals of New York Academy of Sciences*, 916:,61-70.

Salzberg S., C. Kingsford, G. Cattoli, D. Spiro , D. Janies , et al., 2007, Genome analysis linking recent European and African influenza (H5N1) viruses, *Emerging Infectious Diseases*, 13:713-718.

Sasidharan, R., and M. Gerstein, 2008, Protein fossils live on as RNA, *Nature*, 453:729-731.

Savante J., D. Knuth, T. Luczak, and B. Pittel, 1993, The birth of the giant component, arXiv:math.PR/9310236v1.

Sayyed-Ahmad, A., K. Tuncay, and P. Ortoleva, 2007, Transcriptional regulatory network refinement and quantification through kinetic modeling, gene expression microarray data and information theory, *BMC Bioinformatics*, 8:20.

Schapiro S., P. Nehete, J. Perlman, M. Bloomsmith and K. Sastry, 1998, Effects of dominance status and environmental enrichment on cell-mediated immunity in rhesus macaques, *Applied Animal Behavioral Science*, 56:319-332.

Schoepf B.G., C. Schoepf, and J.V. Millen, 2000, Theoretical therapies, remote remedies: SAPs and the politcal ecology of poverty and health in Africa. In: J.Y. Kim, J.V. Millen, A. Irwin, and J Gershman (eds.), *Dying for Growth: Global Inequality and the Health of the Poor*, Common Courage Press, Monroe, ME, p. 91.

Seto, K., R. Kaurmann, and C. Woodcock, 2000, Landsat reveals China's farmland reservers, but they are vanishing fast, *Nature*, 406:121.

Shannon C., and W. Weaver, 1949, *The Mathematical Theory of Communication*, University of Illinois Press, Chicago, IL.

Shi, L., 1993, Health care in China: a rural-urban comparison after the socioeconomic reforms, *Bulletin of the World Health Organization*, 71:723-736.

Shirkov D., and V. Kovalev, 2001, The Bogoliubov renormalization group and solution symmetry in mathematical physics, *Physics Reports*, 352:219-249.

Shortridge, K., 1982, Avian influenza A viruses of southern China and Hong Kong: ecological aspects and implications for man, *Bulletin of the World Health Organization*, 60:129-135.

Shortridge, K., 2003, Avian influenza viruses in Hong Kong: zoonotic considerations, In: Schrijver, R. and G. Koch (eds.), *Proceedings of the Fron-*

tis Workshop on Avian Influenza: Prevention and Control Wageningen, The Netherlands, pp. 9-18.

Shortridge, K., and C. Stuart-Harris, 1982, An influenza epicentre? *The Lancet* xxx:812-813.

Simon V. et al., 2003, Infectivity and replication capacity of drug-resistant HIV 1 variants isolated during primary infection, *Journal of Virology*, 77:7736-7745.

Simpson, J., Y. Shi, O. Li, W. Chen, and S. Liu, 1999, Pig, broiler and laying hen fram structure in China, 1996. Proposal to IARTC International Symposium, June 25-26.
http://sumner.ucdavis.edu/facultydocs/Sumner/iatrc/simpson.pdf.

Singer, P., 2005, Who pays for bird flu? Commentary available at http://www.project-syndicate.org/commentary/singer5.

Sit. V., 2004, China's WTO accession and its impact on Hong Kong-Guangdong cooperation, *Asian Survey*, 44:815-835.

Skierski M., A. Grundland and J. Tuszynski, 1989, Analysis of the three-dimensional time-dependent Landau-Ginzburg equation and its solutions, *Journal of Physics A*(Math. Gen.) 22:3789-3808.

Skurnick J., C. Kennedy, G. Perez, J. Abrams, S. Vermund, T. Denny, T. Wright, M. Quinones and D. Louria, 1998, Behavioral and demographic risk factors for transmission of Human Immunodeficiency Virus Type 1 in heterosexual couples: report from the heterosexual HIV Transmission Study, *Clinical Infectious Diseases*, 26:855-864.

Smith, G., X. Fan, J. Wang, K. Li, K. Qin, J. Zhang, et al., 2006, Emergence and predominance of an H5N1 influenza variant in China, *Proceedings of the National Academy of Sciences*, 103:16936-16941.

Songserm, T., R. Jam-on, N. Sae-Heng, N. Meemak, D. Hulse-Post, et al., 2006, Domestic ducks and H5N1 influenza epidemic, Thailand, *Emerging Infectious Diseases*, 12:575-581.

Soyer O., M. Salathe, and S. Bonhoeffer, 2006, Signal transduction networks: Topology, response and biochemical processes, *Journal of Theoretical Biology*, 238:416-425.

Stefanski, V., G. Knopf and S. Schulz, 2001, Long-term colony housing of Long Evans rats: immunological, hormonal, and behavioral consequences, *Journal of Neuroimmunology*, 114:122-130.

Stewart I., M. Golubitsky, and M. Pivato, 2003, Symmetry groupoids and patterns of synchrony in coupled cell networks, *SIAM Journal of Applied Dynamical Systems*, 2:609-646.

Stewart I., 2004, Networking opportunity, *Nature*, 427:601-604.

Striffler, S., 2005, *Chicken: The Dangerous Transformation of America's Favorite Food*, Yale University Press, New Haven.

Suarez, D., C. Lee, and D. Swayne, 2006, Avian influenza vaccination in North America: strategies and difficulties, *Developmental Biology*(Basel) 124:117-124.

Sun, A., Z. Shi, Y. Huang, and S. Liang, 2007, Development of out-of-season laying in geese and its impact on the goose industry in Guangdong Province, China, *World's Poultry Science Journal*, 63:481-490.

Swetina, J., and P. Schuster, 1982, Self-replication with error – a model for polynucleotide replication, *Biophysical Chemistry*, 16:329-340.

Szathmary, E., and L. Demeter, 1987, Group selection of early replicators and the origin of life, *Journal of Theoretical Biology*, 128:463-486.

Szathmary, E., 1989, The integration of the earliest genetic information, *Tree*, 4:200-204.

Szathmary, E., and J. Maynard Smith, 1995, The major evolutionary transitions, *Nature*, 374:227-232.

Szathmary, E., 2006, The origin of replicators and reproducers, *Transactions of the Royal Society, B*, 361:1761-1776.

Tan, K., and H. Kohr, 2006, China's changing economic structure and implications for regional patterns of trade, protection, and integration, *China and World Economy*, 14:1-19.

Tang, X., G. Tian, J. Zhao, and K. Zhou, 1998, Isolation and characterization of prevalent strains of avian influenza viruses in China, Chinese Journal of Animal and Poultry Infectious Disease, 20:1-5(In Chinese).

Task Force, 1978, Press release dated 27 January, 1978, Office of the Republican Leader of the New York State Assembly.

Tauber A., 1998, Conceptual shifts in immunology: Comments on the 'two-way paradigm', *Theoretical Medicine and Bioethics*, 19:457-473.

Tenallion O., F. Taddei, M. Radman, and I. Matic, 2001, Second order selection in bacterial evolution: selection acting on mutation and recombination rates in the course of adaptation, *Research in Microbiology*, 152:11-16.

Thaler D., 1999, Hereditary stability and variation in evolution and development, *Evolution and Development*, 1:113-122.

Tseng, W., and H. Zebregs, 2003, Foreign direct investment in China: some lessons for other countries. In Tseng, W., and M. Rodlauer (eds.), *China, Competing in the Global Economy*, the Internaltional Monetary Fund, Washington, DC. ,pp. 68-88.

Tucker, J., G. Henderson, T. Wang, Y. Huang, W. Parrish, et al., 2005, Surplus men, sex work, and the spread of HIV in China, *AIDS*, 19:539-547.

UNCHCR 1993, Fact Sheet 25: forced evictions and human rights. http://unhchr.ch/html/menu6/fs25.htm.

Van Eerden, M., R. Drent, J. Stahl, and J. Bakker, 2005, Connecting seas: western Palearctic continental flyway for water birds in the perspective of changing land use and climate, *Global Change Biology*, 11:894-908.

Van Regenmortel M.H.V. and D.L. Hull (eds), 2002, *Promises and Limits of Reductionism in the Biomedical Sciences*, John Wiley Sons, Ltd., Chichester.

Van Valen, L., 1973, A new evolutionary law, *Evolutionary Theory*, 1:1-30.

Villarreal L.P., V.R. Defilippis, and K.A. Gottlieb, 2000, Acute and persistent viral life strategies and their relationship to emerging diseases, *Virology*, 272:1-6.

Volney, W., and R. Fleming, 2007, Spruce budworm (*Choristoneura* spp.) biotype reactions to forest and climate characteristics, *Global Change Biology*, 13:1630-1643.

Waddington C., 1972, Epilogue, in C. Waddington (ed.), *Towards a Theoretical Biology: Essays*, Aldine-Atherton, Chicago.

Wallace D. and R. Wallace, 1998, *A Plague on Your Houses: How New York Was Burned Down and National Public Health Crumbled*, Verso, New York.

Wallace D., and R. Wallace, 2003, The recent tuberculosis epidemic in New York City: Warning from the de-developing world, in M. Gandy and A. Zumla (eds.), *The Return of the White Plague*, Verso Press, New York.

Wallace D., 2001, Discriminatory public policies and the New York City tuberculosis epidemic, 1975-1993, *Microbes and Infection*, 3:515-524.

Wallace R., M. Fullilove and A. Flisher, 1996, AIDS, violence and behavioral coding: information theory, risk behavior and dynamic process on core-group sociogeographic networks, *Social Science and Medicine*, 43:339-352.

Wallace R., D. Wallace and H. Andrews, 1997, AIDS, tuberculosis, violent crime and low birthweight in eight US metropolitan regions: public policy, stochastic resonance, and the regional diffusion of inner-city markers, *Environment and Planning A*, 29:525-555.

Wallace R., and D. Wallace, 1997a, The destruction of US minority urban communities and the resurgence of tuberculosis: ecosystem dynamics of the white plague in the de-developing world, *Environment and Planning A*, 29:269-291.

Wallace R. and D. Wallace, 1997b, Resilience and persistence of the synergism of plagues: stochastic resonance and the ecology of disease, disorder and disinvestment in US urban neighborhoods, *Environment and Planning A*, 29:789-804.

Wallace R., and D. Wallace, 1997c, Community marginalization and the diffusion of disease and disorder in the United States, *British Medical Journal*, 314:1341-1347.

Wallace R., D. Wallace and H. Andrews, 1997, AIDS, tuberculosis, violent crime, and low birthweight in eight US metropolitan areas: public policy, stochastic resonance, and the regional diffusion of inner-city markers, *Environment and Planning A*, 29:525-555.

Wallace R., Y. Huang, P. Gould and D. Wallace, 1997, The hierarchical diffusion of AIDS and violent crime among US metropolitan regions: inner-city decay, stochastic resonance and reversal of the mortality transition, *Social Science and Medicine*, 44:935-947.

Wallace R. and R.G. Wallace, 1998, Information theory, scaling laws and the thermodynamics of evolution, *Journal of Theoretical Biology*, 192:545-559.

Wallace R., and R.G. Wallace, 1999, Organisms, organizations, and interactions: an information theory approach to biocultural evolution, *BioSystems*, 51:101-119.

Wallace R. and R. Fullilove, 1999, Why simple regression models work so well describing risk behaviors in the US, *Environment and Planning A*, 31:719-734.

Wallace R., J. Ullmann, D. Wallace and H. Andrews, 1999, Deindustrialization, inner city decay and the hierarchical diffusion of AIDS in the US, *Environment and Planning A*, 31:113-139.

Wallace R., 1999, Emerging infections and nested Martingales: the entrainment of affluent populations into the disease ecology of marginalization, *Environment and Planning A*, 31:1787-1803.

Wallace R. and R.G. Wallace, 2002, Immune cognition and vaccine strategy: beyond genomics, *Microbes and Infection*, 4:521-527.

Wallace R., D. Wallace, and R.G. Wallace, 2003, Toward cultural oncology: the evolutionary information dynamics of cancer, *Open Systems and Information Dynamics*, 10:159-181.

Wallace R., D. Wallace and R.G. Wallace, 2004, Biological limits to reduction in rates of coronary heart disease: a punctuated equilibrium approach to immune cognition, chronic inflammation, and pathogenic social hierarchy, *Journal of the National Medical Association*, 96:609-619.

Wallace R. and D. Wallace, 2004, Structured psychosocial stress and therapeutic failure, *Journal of Biological Systems*, 12:335-369.

Wallace R., and R.G. Wallace, 2004, Adaptive chronic infection, structured stress, and medical magic bullets: do reductionist cures select for holistic diseases?, *BioSystems*, 77:93-108.

Wallace R., and K. McCarthy, 2007, The unstable public health ecology of the New York Metropolitan Region: implications for accelerated national spread of emerging infection. *Environment and Planning A*, 39:1181-1192.

Wallace R., D. Wallace, J. Ahern, and S. Galea, 2007, A failure of resilience: estimating response of New York City's public health ecosystem to sudden disaster. *Health and Place*, 13:545-550.

Wallace R., and R.G. Wallace, 2008, *Psychopathica Automatorum: A cognitive neuroscience perspective on highly parallel computation and its dysfunctions*. To appear.

Wallace R., and M. Fullilove, 2008, *Collective Consciousness and its Discontents: Institutional Distributed Cognition, Racial Policy and Public Health in the United States*, Springer, New York.

Wallace R., and D. Wallace, 2008, Punctuated equilibrium in statistical models of generalized coevolutionary resilience: How sudden ecosystem transitions can entrain both phenotype expression and Darwinian selection, *Transactions on Computational Systems Biology, IX*, LNBI 5121:23-85.

Wallace, R., and R.G. Wallace, 2008, On the spectrum of prebiotic chemical systems: an information theory treatment of Eigen's paradox, *Origins of Life and Evolution of Biospheres*, DOI 10.1007/s11084-008-9146-1.

Wallace R., 1988, A synergism of plagues: 'Planned Shrinkage', contagious housing destruction, and AIDS in the Bronx, *Environmental Research*, 47:1-33.

Wallace R., 1990, Urban desertification, public health and public order: 'planned shrinkage', violent death, substance abuse and AIDS in the Bronx, *Social Science and Medicine*, 31:801-813.

Wallace R., 2000, Language and coherent neural amplification in hierarchical systems: renormalization and the dual information source of a generalized stochastic resonance, *International Journal of Bifurcation and Chaos*, 10:493-502.

Wallace R., 2002a, Immune cognition and vaccine strategy: pathogenic challenge and ecological resilience, *Open Systems and Information Dynamics*, 9:51-83.

Wallace R., 2002b, Adaptation, punctuation and rate distortion: non-cognitive 'learning plateaus' in evolutionary process, *Acta Biotheoretica*, 50:101-116.

Wallace R., 2005a *Consciousness: A Mathematical Treatment of the Global Neuronal Workspace Model*, Springer, New York.

Wallace R., 2005b, A global workspace perspective on mental disorders, *Theoretical Biology and Medical Modelling*, 2:49.

Wallace R., 2007, Culture and inattentional blindness: a global workspace perspective, *Journal of Theoretical Biology*, 245:378-390.

Wallace R., 2008a, Toward formal models of biologically inspired, highly parallel machine cognition, *International Journal of Parallel, Emergent and Distributed Systems*, DOI: 10.1080/17445760801932357.

Wallace, R., 2008b, Developmental disorders as pathological resilience domains, *Ecology and Society*, 13(1):29.

Wallace R.G., 2003, AIDS in the HAART era: New York's heterogeneous geography, *Social Science and Medicine*, 56:1155-1171.

Wallace R.G., 2004, Projecting the impact of HAART on the evolution of HIV's life history, *Ecological Modelling*, 176:227-253.

Wallace R.G., H. HoDac, R. Lathrop, and W. Fitch, 2007, A statistical phylogeography of influenza A H5N1, *Proceedings of the National Academy of Sciences*, 104:4473-4478.

Wallace, R.G., 2007, The great bird flu name game. H5N1 Blog. http://crofsblogs.typepad.com/h5n1/2007/12/should-we-play.html.

Wallace, R.G., H. Hodac, R. Lathrop, and W. Fitch, 2007, A statistical phylogeography of influenza H5N1, *Proceedings of the National Academy of Sciences*, 104:4473-4478.

Wallace R.G., and W. Fitch, 2008, Influenza A H5N1 immigration is filtered out at some international borders, *PLOSone*, 3(2):e1697.

Wan, X., T. Ren, K. Luo, M. Liao, G. Zhang, et al., 2005, Genetic characterization of H5N1 avian influenza viruses isolated in Southern China during the 2003-04 avian influenza outbreaks, *Archives of Virology*, 150:1257-1266.

Wang, J., D. Vijaykrishna, L. Duan, J. Bahl, J. Zhang, et al., 2008, Identification of the progenitors of Indonesian and Vietnamese avian influenza A (H5N1) viruses from southern China, *Journal of Virology*, 82:3405-3414.

Weber, B., and D. Depew (eds.), 2001, *Evolution and Learning: The Baldwin Effect Reconsidered*, The MIT Press, Cambridge, MA.

Webster, R., M. Peiris, H. Chen, and Y. Guan, 2006, H5N1 outbreaks and enzootic influenza, *Emerging Infectious Diseases*, 12:3-8.

Weinstein A., 1996, Groupoids: unifying internal and external symmetry, *Notices of the American Mathematical Association*, 43:744-752.

Weis, T., 2007, *The Global Food Economy: The Battle for the Future of Farming*, Zed Books, London.

West-Eberhard, M., 2003, *Developmental Plasticity and Evolution*, Oxford University Press, New York.

West-Eberhard, M., 2005, Developmental plasticity and the origin of species differences, *Proceedings of the National Academy of Sciences*, 102:6543-6549.

Whalley, J., and X. Xin, 2006, China's FDI and non-FDI economies and the sustainability of future high Chinese growth, National Bureau of Economic Research, Working Paper 12249.

http://nber.org/papers/w12249.

Whitham T., et al., 2006, A framework for community and ecosystem genetics: from genes to ecosystems, *nature reviews:genetics*, 7:510-523.

Wilson, K., 1971, Renormalization group and critical phenomena. I Renormalization group and the Kadanoff scaling picture, *Physical Review B*, 4:3174-3183.

Wimsatt W.C., 1980, Reductionist research strategies and their biases in the units of selection controversy. In: T. Nickles (ed.), *Scientific Discovery-Volume II: Case Studies*, pp 213-259. Reidel, Dordrecht.

Wolpert D., and W. Macready, 1995, No free lunch theorems for search, Santa Fe Institute, SFI-TR-02-010.

Wolpert D., and W. Macready, 1997, No free lunch theorems for optimization, *IEEE Transactions on Evolutionary Computation*, 1:67-82.

World Health Organization, 2005, Avian Influenza: Assessing the Pandemic Threat.

www.who.int/csr/disease/influenza/WHO$_C$DS$_2$005$_2$9/en/.

Wymer C., 1997, Structural nonlinear continuous-time models in econometrics, *Macroeconomic Dynamics*, 1:518-548.

Xu, K., G. Smith, J. Bahl, L. Duan, H. Tai, et al., 2007, The genesis and evolution of H9N2 influenza viruses in poultry from southern China, 2000 to 2005, *Journal of Virology*, 81:10389-10401.

Xueqiang, X., R. Yin-Wang Kwok, L. Li, and X. Yan, 1995, Production change in Guangdong. In Yin-Wang, R., and A. So (eds.), *The Hong Kong-Guangdong Link: Partnership in Flux*, M.E. Sharpe, Inc., New York., pp. 135-162.

Yang, Y., M. Halloran, J. Sugimoto, I. Longini, 2007, Detecting human-to-human transmission of avian influenza A (H5N1), *Emerging Infectious Diseases*, 13:1348-1353.

Yaron, Y., Y. Hadad, and A. Cahaner, 2004, Heat tolerance in featherless broilers, World Poultry Congress, June 8-12, Istanbul, CD Proceedings.

Yeung, F., 2008, Goldman Sachs pays $ 300 USm for poultry farms, *South China Morning Post*, August 4, 2008.

Yuen, K., and S. Wong, 2005, Human infection by avian influenza A H5N1, *Hong Kong Medical Journal*, 11:189-199.

Zhu R., A. Riberio, D. Salahub, and S. Kauffman, 2007, Studying genetic regulatory networks at the molecular level: Delayed reaction stochastic models, *Journal of Theoretical Biology*, 246:725-745.

Zurek W., 1985, Cosomological experiments in superfluid helium? *Nature*, 317:505-508.

Zurek W., 1996, Shards of broken symmetry, *Nature*, 382:296-298.

Zweig, D., 1991, Internationalizing China's coutnryside: the political economy of exports from rural industry, *The China Quarterly*, 128:716-741.

Index